Oriental Networks

Aperçus: Histories Texts Cultures

Editor: Kat Lecky, Bucknell University

Aperçus is a series of books exploring the connections among historiography, culture, and textual representation in various disciplines. Revisionist in intention, *Aperçus* seeks monographs as well as guest-edited multiauthored volumes, that stage critical interventions to open up new possibilities for interrogating how systems of knowledge production operate at the intersections of individual and collective thought.

We are particularly interested in medieval, Renaissance, early modern, and Restoration texts and contexts. Areas of focus include premodern conceptions and theorizations of race, ethnicity, gender, and sexuality in art, literature, historical artefacts, medical and scientific works, political tracts, and religious texts; negotiations between local, national, and imperial intellectual spheres; the cultures, literatures, and politics of the excluded and marginalized; print history and the history of the book; the medical humanities; and the cross-pollination of humanistic and scientific modes of inquiry.

We welcome projects by early-career scholars; we will not consider unrevised dissertations. Please send a proposal and letter of inquiry to Professor Katarzyna Lecky at kat.lecky@bucknell.edu.

For a full list of *Aperçus* titles, please visit our website at www.bucknelluniversitypress.org.

Oriental Networks

~

*Culture, Commerce, and
Communication in the
Long Eighteenth Century*

EDITED BY BÄRBEL CZENNIA
AND GREG CLINGHAM

Lewisburg, Pennsylvania

Library of Congress Cataloging-in-Publication Data

Names: Czennia, Bärbel, 1961– editor. | Clingham, Greg, editor.
Title: Oriental networks : culture, commerce, and communication in the long
eighteenth century / edited by Bärbel Czennia and Greg Clingham.
Description: Lewisburg, Pennsylvania : Bucknell University Press, 2020. |
Series: Aperçus: histories texts cultures | Includes bibliographical
references and index.
Identifiers: LCCN 2020008425 (print) | LCCN 2020008426 (ebook) |
ISBN 9781684482719 (paperback) | ISBN 9781684482726 (hardcover) |
ISBN 9781684482733 (epub) | ISBN 9781684482740 (mobi) |
ISBN 9781684482757 (pdf)
Subjects: LCSH: East and West—History—18th century. | Globalization—
History—18th century. | Civilization, Modern—18th century. | Social networks—
History—18th century. | Enlightenment—History—18th century.
Classification: LCC CB251 .O724 2020 (print) | LCC CB251 (ebook) |
DDC 909/.09821—dc23
LC record available at https://lccn.loc.gov/2020008425
LC ebook record available at https://lccn.loc.gov/2020008426

A British Cataloging-in-Publication record for this book is available
from the British Library.

♾ The paper used in this publication meets the requirements of the
American National Standard for Information Sciences—Permanence
of Paper for Printed Library Materials, ANSI Z39.48-1992.

www.bucknelluniversitypress.org

Distributed worldwide by Rutgers University Press

Manufactured in the United States of America

Contents

List of Illustrations

Oriental Networks

Introduction

Oriental Networks in the Long Eighteenth Century

BÄRBEL CZENNIA

Contexts

The network has become a cultural icon of the early twenty-first century. Ubiquitous and, in contrast to the object of a net, metaphorical, the term "network" has been used to describe the structure of natural, technological, and social systems. Fanning out from departments of physics and computer science to the social sciences[1] and on to the humanities,[2] network models have been employed to better understand the emergence of new ideas and collective behaviors. Bestsellers such as Albert-László Barabási's *Linked: The New Science of Networks* (2002) and *Bursts: The Hidden Patterns behind Everything We Do* (2010) have popularized network theory[3] and suggested that we live in a world where "Everything is Connected to Everything Else" and where networks form the basis of "Business, Science, and Everyday Life."[4] Bruno Latour has tried to explain the success of Louis Pasteur's revolutionary theory of germs in nineteenth-century France and the failure of a modern urban transit system in contemporary Paris with the help of "actor-network theory."[5] Attracting both praise *and* controversy,[6] network-centered approaches have also served to analyze the complex relations between modern citizens of the world who find themselves, for better and for worse, hyperconnected through communication networks, transport networks, social networks, even metabolic networks and networks of disease.

The roots of the modern "Age of Networks"[7] reach back to the early modern period, a time associated with rapid improvement of transportation technologies, the expansion of intercultural contacts, and the emergence of a global economy. Then as now, global connectivity generated new opportunities and new challenges, reshaping every aspect of human interaction. Then as now, metaphors employed to describe new types of circulatory and connective configurations on a global scale reflected both the hopes and the unease experienced by people affected by such changes.

Oriental Networks explores forms of interconnectedness between the Western and the Eastern hemispheres during the long eighteenth century. In eight case studies, it examines relationships between individuals and (informal or formal) groups that could be regarded as precursors to modern networks and whose participants engaged in forms of intercultural exchange that might qualify as networking in the modern sense. Contributors ask what sorts of practices networking involved, how and where the exchange processes took place, and what effect they had on individual and collective actors previously separated by vast geographical distances and cultural boundaries. Networks are sometimes favorably compared to other forms of organization as less deterministic or hierarchical.[8] Rather than idealizing networks as inherently superior, however, this essay collection also considers eighteenth-century expressions of resistance to networking that preceded modern skepticism toward the concept of the network and its politics.[9]

Whether employed by sociologists to rewrite the history of science and technology or by counterintelligence officers to fight international terrorism, network studies focus on phenomena of growth and expansion involving cooperation between great numbers of people and institutions.[10] Three central assumptions shared across disciplines and, therefore, also pertinent to the study of networking activities during the eighteenth century, include the idea that "many interacting agents following simple rules can collectively produce complex behaviours,"[11] that networks are dynamic structures, constantly evolving and adapting to changing environments, and that network-centered approaches are ideally suited to understand complex processes of innovation and the creation of new knowledge. As I suggest below, eighteenth-century citizens were fully aware of living in an increasingly complex world of many interconnected actors, quickly changing environments, and accelerated innovation processes.

The network has always been a highly ambivalent image with positive and negative connotations, ranging from joyful participation to dangerous entrapment. On the positive side, Joseph Addison's *Spectator* 69 celebrated merchants as the most "useful Members in a Commonwealth" because they "knit Mankind together in a mutual Intercourse of good Offices, distribute the Gifts of Nature, find Work for the Poor, add Wealth to the Rich, and Magnificence to the Great."[12] Not unlike modern career coaches who recommend networking as a survival skill for young professionals entering a competitive world market, Oliver Goldsmith's London-based Chinese philosopher Lien Chi Altangi declared it to be "the duty of the learned to unite society more closely, and to persuade men to become citizens of the world."[13] The presentation of global connectivity as a must in the modern business world finds its precursor in John Dyer's poem *The Fleece* (1757), where the author sketches the optimistic vision of a worldwide "woolen web" (3:483) emanating from the early industrial English Midlands or "webs of Leeds" (4:142). Not only does the poet anticipate the key metaphor of the digital age, the network, by several hundred years, Dyer's emulation of the ancient Georgic also celebrates Britain's growing economic linkage with distant parts of the world as an act of philanthropy, enmeshing humans and even entire coastlines of foreign continents with a warming layer, woven on British looms:

> . . . a day will come,
> When, through new channels sailing, we shall clothe
> The Californian coast, and all the realms
> That stretch from Anian's streights to proud Japan.[14]

All three eighteenth-century authors imagine human professionals (merchants and members of the educated elite, possibly including diplomats and politicians) either as moving through networks of transport and commerce or as networkers (human "nodes") facilitating and expediting the exchange of objects and information around the world.

Private individuals of the present day who satisfy their material, informational, and even emotional needs by connecting themselves to the Internet would have appealed as much to Addison's Mr. Spectator as to Goldsmith's Asian alter ego, both of whom associated global trade and consumerism with social stability, economic prosperity, and world peace. In Addison's ontological worldview, the irregular distribution of natural resources became a means to heighten mankind's awareness of their "Dependance upon one another" and of being "united together by their common Interest" in "mutual Intercourse and Traffick."[15] Addison's emphasis on the unifying effect of global commerce is echoed in Goldsmith's postulate that "[T]he greater the luxuries of every country, the more closely, politically speaking, is that country united. Luxury is the child of society alone; the luxurious man stands in need of a thousand different artists to furnish out his happiness; it is more likely, therefore, that he should be a good citizen, who is connected by motives of self-interest with so many, than the abstemious man who is united to none."[16] The close nexus between commercial and social networking clearly is not a modern invention.

The long eighteenth century also anticipated modern ambiguities in the usage of the term "network," which could be applied to structures consisting either of people or of non-human entities—for instance, cities and harbors within a transport network. Long before the arrival of modern highspeed data transmission through computer networks and of world travel through interconnected aviation hubs, poets condensed the emerging transport network of global sea routes into the now familiar imagery suggestive of nodes and edges. Fifty years after John Dyer envisioned British-controlled harbors and waterways ("channels" and "canals") as building blocks of an ever-expanding transport network for the worldwide distribution of British wool, William Wordsworth repurposed the underlying idea of global circulation through interconnected tubular structures for yet another poetic image celebrating Britain's imperial expansion. An extended metaphor in his sonnet "Composed upon Westminster Bridge, September 3, 1802" presents London as a "mighty heart" pumping vital nutrients through the circulatory system of an imperial body—a physiological network of sorts—and thereby legitimizes the political entity of an empire as organic, natural growth.[17]

While all examples offered so far illustrate perceptions of global connectivity as benign and life-enhancing, eighteenth-century prefigurations of modern networks were by no means unanimously welcomed as positive developments. Then as now, critical voices also exposed negative side effects of networking, especially the damaging impact of global trade on the social fabric of local communities. Just as social justice movements today draw attention to the collateral social, political, and environmental damage caused by neoliberal globalization, Oliver Goldsmith's poem "The Deserted Village" (1770) lamented the destruction of traditional ways of life and the special vulnerability of the rural poor ensuing from processes of modernization, especially from the quick growth of "trade's proud empire."[18]

Even Adam Smith, although a strong advocate of Enlightenment belief in political and economic progress, was "deeply ambivalent about globalization" and exposed international trading companies of his day as threats to the independence of European governments and as ruthless oppressors of local economies in Asia and other continents.[19] Although he considered the "discovery of America, and that of a passage to the East Indies by the Cape of Good Hope" to be "the two . . . most important events recorded in the history of mankind," Smith grimly concluded:

> By uniting, in some measure, the most distant parts of the world, by enabling them to relieve one another's wants, to increase one another's enjoyments, and to encourage one

another's industry, their general tendency would seem beneficial. To the natives, however, both of the East and West Indies, all the commercial benefits which can have resulted from those events have been sunk and lost in the dreadful misfortunes which they have occasioned.[20]

Not only did Smith accuse transnational social, economic, and political institutions of the eighteenth century as having "entirely neglected" the interests of "consumers" at home and abroad; by apostrophizing the responsible parties as "this whole mercantile system,"[21] he also came very close to identifying (in all but name) what modern writers would define as imperial networks.[22]

Adam Smith's disillusionment with the systemic cruelty and corruption of transnational networks was shared by Anna Letitia Barbauld, who cast her reservations about global connectivity in verse rather than in prose. Not unlike modern psychologists who worry about the *mental* health of people no longer able to disentangle themselves from the World Wide Web or modern media who discuss digital addiction as a potential threat for future generations, Barbauld expressed concern for the *moral* health of her compatriots in the "Epistle to William Wilberforce" (1791). Rather than with the circulation of commerce and wealth, she associated Britain's participation in the transatlantic slave trade with the global circulation of disease and corruption:

> Corruption follows with gigantic stride,
> And scarce vouchsafes his shameless front to hide:
> The spreading leprosy taints ev'ry part,
> Infects each limb, and sickens at the heart.[23]

Barbauld also exposed the damaging effect of British "Avarice" (l. 25) on the physical integrity of individual slaves—evidenced by "deep groans," "constant tear," and "bloody scourge," (ll. 5–8)—as well as on the political stability of entire African communities, likewise visualized as injured bodies. Torn apart by intertribal warfare of "[c]ontending chiefs" (l. 23) eager to make prisoners in order to meet European nations' rising demand for slaves, "[s]till Afric bleeds, / Uncheck'd, the human traffic still proceeds" (ll. 15–16), as long as "Britain's senate" (l. 39) participates in the lucrative but unethical triangular trading system that links a multitude of players in Europe, West Africa, and European colonies in the Americas, the Caribbean, and British North America.

Barbauld's "Epistle" suggests that personal and institutional, economic and political entanglements of the transatlantic slave trade as well as many other global exchanges that occurred in the context of Britain's empire building to the east (referenced in the last part of her "Epistle") involved a higher degree of complexity than earlier forms of intercultural exchange. Developing unforeseen dynamics of their own and unexpected results, these multilayered, mutual dependencies already resembled the more complex relationships that modern researchers define as networks.[24] The long eighteenth century appears to be the period in history when less complex forms of connectivity evolved into full-fledged networking on a global scale, thereby intensifying a development that Adam Smith had traced back to the European discovery of America:

> By opening a new and inexhaustible market to all the commodities of Europe, [the discovery of America] gave occasion to new divisions of labour and improvements of art, *which, in the narrow circle of the antient commerce, could never have taken place for want of a market* to take off the greater part of their produce. . . . The commodities of Europe were almost all new to America, and many of those of America were new to Europe. A new sett [*sic*] of exchanges . . . began to take place which had never been thought of before, and which should naturally have proved as advantageous to the new, as it certainly did to the old continent. The savage injustice of the Europeans rendered an event, which ought to have been beneficial to all, ruinous and destructive to several of those unfortunate countries.[25]

Historians have long focused on the highest level of government as the driving force to initiate and to steer processes of exchange across cultural divides and vast geographical distances. Recent approaches have redirected attention from macrosocial factors to the many interpersonal levels along the way at which intercultural exchanges were negotiated and maintained. Network-centered approaches complement and complicate traditional historical accounts by highlighting the overlay of local alliances, private groups, and public institutions involved in such globe-spanning undertakings and by accommodating the many unintended and often unexpected outcomes. The number of actors operating within and across networks, and across long distances was too high for any single organizational unit to be solely and fully in control. Social, economic, and imperial historiographers are currently replacing traditional master narratives with more "relational" approaches that emphasize "local specifics" and the "complexity of dynamics" involved in transnational and supra-regional exchanges.[26]

Eighteenth-century scholars have begun to explore networks such as literary clubs, salons and related types of social connectivity, regional economic networks, and urban communication networks.[27] Even before the recent uptake in network studies, researchers had been aware of the global scope of the Enlightenment, as is evident from the scholarship highlighting worldwide exchange.[28] This global turn also resulted in a reassessment of the impact of the East on the cultural imagination of the West.[29] The current volume combines the global perspective with a special focus on the Far East and traces the complex web of intersecting networks that linked East and West. It contributes to our knowledge of intercultural exchange between what was once called the "Orient" and the Western world.[30] Contributors to the volume ask how networks of transport, trade, politics, social life, knowledge, and (verbal or visual) art facilitated, affected, and changed relations between East and West across extensive geographical, political, religious, ethnic, and linguistic boundaries over a period of roughly 150 years (circa 1660 to 1840).

In the emerging interdisciplinary field of network-centered cultural history, it never hurts to heed the advice of one's predecessors. Following the call of Beattie, Melillo, and O'Gorman for historic specificity[31] and Casson's request for more local detail,[32] *Oriental Networks* contributes to an evolving new perception of relations between East and West. Its case studies explore the relevance of networks for eighteenth-century encounters with "exotic" geographical locations and peoples, as well as for the representation of the East in British poetry, fiction, drama, aesthetics, landscape design, visual art, and material culture.

Scope and Structure

Several essays in this volume focus on relations between eighteenth-century Britain and a relatively distant East, especially East Asia (China) and South Asia (the Indian subcontinent). Others examine networks with nodes that were located in regions closer to the West and are nowadays apostrophized as the "Middle East" (Western Asia with Egypt and the Eastern Mediterranean world of the Levant), a term that is gradually replacing the nearly coextensive "Near East." The decidedly Eurocentric and somewhat oscillating connotations of the term "Orient" conveniently match the comparatively wide geographical scope of the volume. Just as the noun "Orient" and the adjective "oriental" could designate various land masses situated to the east *and* to the south of Europe and, over time, shifted in meaning both with perceived and with objectively changing cultural and territorial frontiers, some essays cover territory in North and Northwestern Africa and Moorish Spain (conflated in the term "Barbary" or "Barbary Coast"). Several essays also touch on the role of the Cape Colony as a pivotal geographical link (or node) in various types of networks that connected the Eastern and Western hemisphere.

On one level, *Oriental Networks* moves from nonhuman entities (plants and animals) to networks facilitating the movement of complex ideas and on to the role of individuals who acted as networkers and disseminated objects and ideas (or enabled such processes). Yet on another level, those three categories cannot be kept entirely separate as soon as one begins to disentangle the overlay of "nested sets"[33] of interconnected networks involved in every intercultural exchange. Nor are the outcomes of such exchanges as predictable as they may seem. Contrary to some theorists' associations of Western imperialism with a consistent pattern of hostile "orientalism,"[34] contributors to this volume show that encounters with Eastern goods, ideas, and people during the eighteenth century did not always or exclusively serve to reinforce a Western sense of cultural superiority. Eastern commodities accessed through interlocking networks of commerce, social life, knowledge, and art did, on the contrary, often have the effect of transforming both the imported products and their Western beneficiaries. While Eastern imports were recharged with new meanings to increase their appeal to Western sensibilities, occidental consumers were in turn subtly orientalized by new rituals and behaviors acquired in the process of adopting Eastern commodities. Western travelers and temporary migrants who spent extended periods of time in the East for personal or professional reasons likewise found themselves culturally transformed with often unexpected and sometimes life-changing consequences. As cultural interlopers, they became bridge builders who served as human "hubs" in eighteenth-century networks of knowledge.

While cultural transfer rarely comes without transformation and also involves reinterpretation or creative appropriation, intercultural brokers contributed significantly to the dissemination of Eastern culture as a source of Western enrichment or even renewal.[35] "Oriental" networks in this sense do not comprise only the land and sea routes, chartered trading companies, merchant navies, and colonial administrations that bridged the geographical distance between the Far East and the Far West[36] but also the many social, intellectual, discursive, and artistic networks that reduced cultural distance and accelerated processes of transculturation.

Transculturation through networking as an overarching theme links the first three essays, all of which focus on the exchange of material goods, including live plants and animals. A preliminary clarification may, therefore, be helpful for the understanding of "transculturation," a term that is used by several contributors and, although not altogether new, "is newly inflected as it is applied to different fields."[37] Originally coined in 1940 by Fernando Ortiz to describe "complex transmutations of culture" in Cuba,[38]

> the word *transculturation* better expresses the different phases of the process of transition from one culture to another because this does not consist merely in acquiring another culture, which is what the English word *acculturation* really implies, but . . . also . . . involves the loss or uprooting of a previous culture. . . . In addition it carries the idea of the consequent creation of new cultural phenomena, which could be called neoculturation.[39]

In contrast to "interculturality," according to Wolfgang Welsch a concept that constitutes different cultures as separate "spheres or islands," and in contrast to "multiculturality," which assumes that even "*within one society*" different cultures coexist as "clearly distinguished" and "in themselves homogeneous" entities, "transculturality sketches a different picture of the relation between cultures. Not one of isolation and conflict, but one of entanglement, intermixing and commonness. It promotes not separation, but exchange and interaction."[40] Emphasizing the "creative potential of cultural encounters," the term "transculturation" looks "beyond the syncretic model of two cultural systems co-existing to embrace those elements retained and lost by the two systems in the creation of a third."[41]

The essays by Richard Coulton, Stephanie Howard-Smith, and me demonstrate that exchanges of material culture between East and West always involved much more than the geographical boundary crossing of goods. Because the imported goods could never be separated from their cultural meanings, they also affected the behavior of transcultural adopters as well as their personal or collective identities. This holds true for the introduction of the first modern global commodity, tea, to Britain (Coulton) as well as for the circulation of live dogs and their porcelain counterparts through networks linking China and Europe (Howard-Smith) and also for the exchange of garden plants and garden aesthetics between eighteenth-century Britain and British India (Czennia). All three contributors show how such exchanges affected recipients at both ends of the exchange process in ways that went far beyond the acquisition of the imported objects themselves. Networked goods modulated local tastes (both in the literal sense of taste buds and in the more comprehensive sense of aesthetic norms); they introduced new topics for artistic representation and resulted in the creation of new types of furniture and collectibles. They also altered daily rituals, leisure-time activities, and even large-scale outdoor environments.

In a double approach that examines the Western adoption of tea both as a beverage and as a plant, Richard Coulton traces intersecting networks of professional plant prospectors, the institutionalized commercial network of the British East India Company, an emerging network of Royal botanic gardens, personal networks of individual merchants, and transcultural networks of knowledge, all of which were involved in transforming tea from an outlandish luxury item, accessible to only a small social elite, into an everyday commodity enjoyed by many. Rather than as a "mere acquisition of culture,"[42] however, Coulton presents the movement of tea between Eastern and Western hemispheres as a "reciprocal process" that effected the cultural connotations of tea itself as much as the geocultural contexts between which it moved. Tea changed Britain (by producing new social routines and cultural codes) and was simultaneously changed by Britain (where it was consumed and understood differently compared with China). At the same time, some of the original (Chinese) meanings and traditions of tea consumption were translated and partly preserved; ultimately, the economies and meanings of tea in China were in part altered as well.[43] By detailing these complexities, Coulton reminds his readers that intercultural networking "far exceeds the simple combining of cultural elements that the idea of 'fusion' might suggest" and that processes of transculturation result in "heterogeneity within specific geographical locations."[44]

As chapters of (inter-) cultural history that, in Coulton's words, were neither "inevitable" nor "the calculated outcome of a single strategic plan," all three essays showcase the benefits of network-centered approaches. Less top-down-oriented and "less exclusively focused on economic and political issues," they emphasize connectivity below the highest levels of administration.[45] Instead, they show how intercultural exchanges often followed their own dynamics, to some extent independent of governments, and recognize that intercultural relations "were always stretched in contingent and non-deterministic ways."[46] By tracing the many individuals whose personal and professional, commercial and scientific networks facilitated the transplantation of *Camellia sinensis* as well as many other Asian plants to the Western hemisphere, Richard Coulton and I also highlight the importance of international and intercultural collaboration for successful exchanges.

The fact that such ambitious joint ventures of Western botanists, gardeners, and imperial administrators would have been impossible without the support of their local agents in Asia has long been overlooked in more traditional, Eurocentric historiography. Coulton's recording of the many trials, errors, and temporary setbacks experienced by intercultural networkers during the long eighteenth century also exposes the fragility of networks at a time when one missing link could easily destroy the endeavors of many years and many hands.[47] The inherent ephemerality of all networks, whether personal or institutionalized, whether botanical, political, or artistic, is also an overarching theme linking Coulton's essay to those by Stephanie Howard-Smith, me, and, later in the volume, also Jennifer Hargrave and Greg Clingham.

In contrast to Coulton, who traces the one-way movement of an oriental commodity from East to West (albeit one that entails alterations of the meanings of tea in both hemispheres), Stephanie Howard-Smith examines *repeated* movements of lapdogs and porcelain figurines back and forth between East and West as a multistep process of transculturation.[48] Live dogs and porcelain dogs circulating between the two hemispheres could almost be called products *of* transculturation per se: subjected to *multiple* rounds of global exchange and successive stages of artful orientalizing along the way, they became so genuinely hybridized as to make it nearly impossible to retrace their origins in a particular part of the world, at least not before the arrival of modern geneticists and academically trained art historians. In the case of the China figurines, Howard-Smith argues, the combining of cultural markers from East and West was by no means accidental but rather motivated by economic ambition of multiple national competitors fighting for shares in a globalized market. In an astounding episode of industrial plagiarism, Chinese copycat versions of orientalized porcelain dogs, originally designed and produced in Germany, were reintroduced to the West to cater to a growing European demand for Chinoiseries. How live pugs came to be associated with China in the first place, how porcelain figurines and live dogs circulated through intersecting networks of commerce and transportation as personal or diplomatic gifts and as commercial cargo, how the porcelain pug rose to momentary fame as one of the most popular collectibles for connoisseurs before its equally abrupt disappearance, all of this forms part of Howard-Smith's account of an unusually complex process of transculturation.

Similar to Coulton's and Howard-Smith's essays, my own exploration of intercultural gardening involves transregional exchanges of bioactive material through overlapping networks of plant collectors, amateur and professional gardeners, imperial administrators, and visual artists. But as it takes more than plants to design eighteenth-century landscape gardens, I also consider the cultural appropriation and adaptation of the underlying aesthetic concepts, thereby expanding another overarching theme of the volume, the limits and constraints of networks and global exchange relationships during the eighteenth century. Transculturation resulting in "depreciation and loss" rather than "transformative acquisition"—first illustrated by Coulton's example of the Europeans' inability to establish the tea plant as a cash crop in the West—to some extent also affected the phenotype of gardenscapes in East and West; but in addition to the human factor, local climate hampered all-out attempts to Anglicize Indian gardens or to Indianize English gardens. In the case of gardens, however, losses (for instance, of authentic plant material) could be counterbalanced by the introduction of new (for instance, architectural) ornaments, which in turn facilitated the introduction of new cultural meanings. As artistic creations, the transculturated gardens that emerged from networking across hemispheres confirm Fernando Ortiz's observation that "the result of every union of cultures is similar to that of the reproductive process between individuals: the offspring always has something of both parents but is always different from each of them."[49] Coulton, Howard-Smith, and I also show how the global circulation of objects and ideas (through networks of transport, commerce, and knowledge) resulted in tangible, innovative cultural products and altered human behaviors.

In the case of gardens, members of the social and educated elite—including aristocrats, scientists, colonial administrators, and artists who shared a private interest in horticulture—often served as multiply connected human nodes at the intersections of formal (public, institutionalized) and informal (private, interpersonal) networks and accelerated processes of transculturation. Interestingly, artists did not only *represent* garden-related objects and networking activities, for instance, in poems, travel writing, and paintings that thematized or visualized Asian plants, gardens, or gardening activities. In a "feedback loop" not untypical of adaptive systems, including complex networks,[50] painters and landscape architects could also become *active performers* within networks of imperial commerce and institutionalized knowledge production, traveling back and forth between East and West and thereby changing the networks themselves as well as the cultural innovations facilitated by global networking activities.

With Samara Cahill's and Jennifer Hargrave's essays, the volume moves from exchanges of tangible objects and aesthetic concepts to more complex belief systems and to the role of human actors in their dissemination across networks of evangelism and knowledge. Both contributors focus on characters involved in oriental missions but examine transregional networks and human commitments that go far beyond the stereotypical activities of eighteenth-century Christian missionaries. Working from very different textual genres—early eighteenth-century episodic novels (Cahill) and Romantic-era didactic dialogues (Hargrave)—both authors explore the spiritual, political, and ethical implications of the engagement in networks for the networkers themselves.

Cahill concentrates on reading novels closely. Her approach is double: on the one hand, it considers literary representations of contemporary networks as constituents of the narrated world. On the other hand, it reflects the inherently figurative language of literary texts and emphasizes the metaphorical uses of networks and their structural functions for verbal art and for the artist's perception and interpretation of historical circumstances. Penelope Aubin's exposure of British entanglement in transcontinental networks of commerce—in her novels *The Strange Adventures of the Count de Vinevil and His Family* (1721), *The Life of Charlotta du Pont, an English Lady* (1723), *The Life and Adventures of the Lady Lucy* (1726), and *The Life and Adventures of Young Count Albertus* (1728)—becomes part of a comprehensive questioning of Western empire building based on ethical concerns. The remedy against Britain's political infighting and material greed is brought about by Christian redeemer figures whose self-sacrifice calls for a radical reorientation of collective values. Aubin contrasts the negatively connotated circulation of British capital through imperial networks of finance with the positively connotated symbolic countercurrency of martyrial blood, which flows from the tortured bodies of Eastbound members of a religious network of Christian believers before it is metaphorically recirculated back to the West through transnational communication networks. What makes Aubin's approach to national unity innovative and unexpectedly modern, despite its outdated melodramatic mode, Cahill argues, is her redefinition of British identity as inherently hybrid. Conceived by culturally composite parents from diverse national and religious affiliations, Aubin's transculturated characters symbolically unite—through the "mixed" blood circulating in their bodies—all the antagonistic global players with whom Britain is linked through economic networks, yet with whom it also competes for political dominance and material gain. By organizing her multistranded plots around verbal images of circular motion suggestive of more altruistic forms of networking, Aubin generates a literary response that criticizes empire building. Not only does this reading of Aubin reconnect an underappreciated literary pioneer to the important literary tradition of literary globalism, Cahill thereby also contributes to the exploration of conflicts and dissonances related to eighteenth-century imperial networking, likewise explored by several other essays, including that of Jennifer Hargrave on Robert Morrison, Protestant missionary to Portuguese Macao, Qing-era Guangdong, and Dutch Malacca, as well as pioneering sinologist, also remembered as the "Father of Anglo-Chinese Literature."

If Aubin expressed her unease with British imperialism through fictional antitrade heroes or heroines and through her rereading of the East as the space of Britain's spiritual redemption and renewal, Robert Morrison inscribed his resistance to British imperialism into forms of writing that have long remained underexplored. By retracing Morrison's innovative contributions to and movements within oriental networks of evangelism, commerce, and knowledge—often regarded as unanimously supportive of Britain's Eastbound imperial expansion—Jennifer Hargrave's essay questions the one-sided portrayal of Morrison as emissary of Western Christianity and enabler of British imperialism. Her rereading of this unconventional cultural mediator confirms the value of network-centered approaches for the correction of simplistic accounts of imperial power structures. While Hargrave does not question Morrison's professional commitment to the London

Missionary Society (as a preacher) and the East India Company (as a translator), she argues that his deepening scholarly interest in Eastern culture—evidenced by his involvement with both Western and local Chinese networks of knowledge—increasingly induced him to work outside, at times even against already established networks.

Accused of underperformance by his official employers, Morrison became a pioneering sinologist who established an alternative coterie of scholars, missionaries, employees of the East India Company, and diplomats that bypassed extant networks, especially those that attempted to fit China within broader European imperial interests.[51] Hargrave reminds us that Morrison's Anglo-Chinese coterie united many individuals who likewise did not fit precisely within the networks they officially represented. Their private collaboration in an informal network that questioned the official imperialist agenda of their employers calls for revision of an older imperial historiography that edited out paradoxes and inconsistencies of empire building in the interest of a homogeneous picture of unilateral agency. Hargrave's preference for the term "exchange" instead of "network" for her characterization of Morrison's intercultural and multicultural coterie draws attention to its informal nature. The fact that intercultural collaborations were officially forbidden by the Chinese government and necessitated covert interactions between its members might even justify its characterization as a sort of "underground" network.

Through his activities as a cofounder of the Anglo-Chinese College in Malacca, translator, and writer, Morrison worked toward a new type of network that was smaller and more personal but never fully actualized until after his death. Hargrave's emphasis on collaboration as an important feature of intercultural networking links her essay to those by Coulton and myself at the beginning of the volume. Hargrave's portrayal of Morrison as network enabler *and* network critic concludes with a discussion of the transformative effect of life in an eighteenth-century contact zone. As a place where processes of transculturation might affect an individual's scholarship just as much as his personal identity, it offers further evidence of the fluidity and reciprocity of relations between actors within networks. Mediating between intersecting networks and between cultures, Morrison morphed from an isolated human node of eastbound Western evangelism into a multidirectionally connected oriental hub facilitating vast flows of scholarly information.

NETWORKERS OF EASTERN EMPIRE

Morrison's innovative take on networking through intercultural dialogue comes very close to Margrit Schulte Beerbühl's definition of a "cosmopolitan network" as one

> whose members not only cross national boundaries through their activities and mobility, but connect distant places and foreign cultures through their activities, and contribute to the generation of knowledge and goods, and to a social and cultural interconnectedness that brings about an understanding of the other as well as a curiosity and interest in engaging with the stranger.[52]

In a remarkable twist of imperial history, the future-oriented cosmopolitanism of Morrison's innovative coterie on the fringes of established imperial networks found its contemporaneous opposite in the demonstrative anticosmopolitanism of Charles Lamb, whom James Watt aptly characterizes as "'networked' by virtue of his occupation." Watt's exploration of Lamb's conflicted identity as low-ranking administrator in the service of the East India Company and as literary author also provides an interesting typological contrast to Greg Clingham's subsequent reconstruction of English diplomat George Macartney's steadfast engagement in the overlapping networks of politics, commerce, and cosmology as Britain's first envoy to China. All three essays suggest that success and failure in the service of empire are at times difficult to distinguish or may even lie entirely in the eye of the beholder.

Watt presents Lamb as a network skeptic in the sense that he attempted "to resist the role which he knew he occupied" through subtle point-of-view oscillations in his essays. This is exemplified by an essay by Lamb that explores "how it might be possible for people in the imperial metropolis to imagine themselves as un-networked" in the sense of "un-connected to the sources of their wealth."[53] Watt's discussion of Lamb as a critic of imperial networking from within the net contributes an important facet to a central theme of this volume, that of eighteenth-century interrogations of networks as immanent critiques of globalization (also explored by Jennifer Hargrave and Samara Cahill). Watt's essay shows how enthusiasm for the imperial agenda could wear thin even close to the imagined "center" of an empire and even during the heyday of Britain's eastward expansion of influence.

At first glimpse, Charles Lamb's self-distancing from the imperial idea through writing in London seems to resemble that of Robert Morrison in faraway Canton. But where Morrison moved from writing to acting (through intercultural mediation and subversive networking), Lamb's critique of empire remains in the realm of rhetorical performance. Watt shows that Lamb's disidentification with his professional network is neither explicitly based on ethical reservations (à la Aubin) nor on some enlightened perception of world cultures as equally worthy members of a global family (à la Morrison). Instead, Lamb's resentment appears to be far more self-centered and mainly fueled by the frustrations of an intellectual unable to market his literary talent in an imperial economy in such a way as to spare him the dependence on a menial breadwinning job in the service of empire. Watt's comparison of Lamb with several intellectuals holding higher-ranking positions as administrators of the East India Company in the first half of his essay confirms the impression of an idiosyncratic situation and of Lamb as a disloyal individual, a weak link, so to speak, in the imperial network that he publicly represented.

Watt's close reading of Lamb's semifictional Elia essays in the second half of his argument, however, reveals far more complex feelings of unease about the imperial agenda that Lamb obviously shared with many of his contemporaries. What is characterized as the "staging" of a new ideological formation could also be regarded as a literary author's soul-searching on behalf of the whole nation. In the process of a narrative dramatization that is cast in the form of fictional Elia's nostalgic remembrance of former colleagues—a personal network of professional connections in the fictional space—Lamb confronts anxieties that flow from a collective sense of shame or guilt. The latter also reveals a suppressed awareness of the dark side of empire—an aspect that the enjoyment of modern amenities of metropolitan life could not fully erase: Britain's entanglement in unsavory practices such as slavery, military aggression, illegal opium trade, and institutionalized racism. Lamb's step-by-step excavation of "submerged links" between metropole and colony gave him the appearance of a networked misfit; but by paving the way for a whole range of complex new literary characters with split identities that mirror collectively experienced cultural unease, so James Watt reminds us, Lamb became an overachiever in the realm of art.

As imperial administrators, both Lamb and Macartney operated in the sphere of rhetorical performance that linked actual (political and economic) networks and their symbolic representations. Whereas Lamb's unease with the East India Company expressed itself in a subversive rhetoric that challenged the public image of the commercial network he officially represented, Macartney's unquestioning identification with the interests of British government and of the East India Company in China should, therefore, have paved the way for a much smoother career as a political networker. In yet another ironic twist of history, however, Macartney's attempt to develop economic relations between Britain and China were hampered by cultural difference, a significant constraining factor for exchange, in addition to technical error (thematized by Coulton) and climatic incompatibility (discussed in my own essay).

In a collection of network-centered case studies that emphasize intercultural connectivity *below* the highest level of government, Macartney forms an interesting typological counterexample for he employed cultural networking strategies in diplomatic negotiations between King George III and the Qianlong emperor.

While contemporaries and later historians have presented Macartney's embassy to China as a failure, Greg Clingham emphasizes Macartney's achievements as an intercultural networker. Not only was Macartney keenly aware of the importance of global commercial and cultural networks already linking eighteenth-century Britain and China; he also developed diplomatic gift-giving into a symbolic sign language that reinterpreted political exchanges within broader cultural and cosmological networks. In his "style of writing and thinking," Clingham argues, the British diplomat displayed an acute awareness of the cultural distance between the Western and the Eastern empires, a distance that he sought to reduce rhetorically by blending the discourses of diplomacy and trade symbolically by framing them with cosmological tropes. Although the incompatibility of Western and Eastern concepts of the universe ultimately frustrated the British purpose in gifting telescopes and orreries to the Chinese, and resulted in a rejection of Macartney's attempted intercultural bridge-building, this diplomatic setback does not in and of itself diminish Macartney's achievement as a perceptive and creative mediator between East and West. Based on evidence gained from the analysis of unpublished diplomatic correspondence and memorandums in the Wason Collection at Cornell University and the Toyo Bunko Oriental Library in Tokyo, Clingham challenges the traditional historiographic verdict and reconstructs Macartney as an example of a more culturally sensitive negotiator. In this reading, Macartney seeks to meet his Asian negotiating partners halfway between the hemispheres by emphasizing commonality rather than distance and by linking the spheres of cosmology and commerce. What we learn from Clingham, then, is that diplomatic rhetoric may become a central vehicle for intercultural communication. What appears on the surface as rhetorical bungling contributes to a history—as yet unwritten—of *attempted* political and economic networking, perhaps even to the invention of diplomacy and cosmological science as a form of cultural networking.

The concluding essay of the volume explores Eastern-Western connectivity from the point of view of an eighteenth-century woman and travel writer who deserves more scholarly attention than she has received so far. Kevin Cope's XXL-sized case study explores the impact of extant global networks on Maria Graham's intercultural encounters, as well as the networks that she herself created along the way. Moreover, it demonstrates how Graham developed networking into a sophisticated narrative technique, eventually also into a new way of looking at a shrinking world that already looked very modern. Cope's double study explores networked material culture, social, and professional networks as well as networking as a comprehensive and innovative narrative mode that was perfectly suited to capture a globalized eighteenth century.

Cope portrays Graham as an open-eyed observer of a globalized economy that enabled competing empires to circulate Eastern commodities into every corner of the discovered world. He also presents her as a social and intercultural networker and, even more interestingly, as a self-conscious semiliterary writer whose perceived and constructed links between East and West developed an unexpected dynamic that affected her narrative technique. Not only did Graham compose narratives *about* extant networks; the artist's narrative web also morphed into a new sort of *networked* narrative. Cope shows how Graham's technique of connecting the dots between far-flung destinations, cultures, and individuals for her armchair-traveling readers turned the author into a "super-node" whose creative mind provided a space where horizontal (geographical) and vertical (historical) filaments of graveyards, vegetable plots, cave systems, and mountain tops intersected.

For Graham, Cope argues, networking was not only a modus operandi that informed her social activities, her travel itineraries, and her narrative mode. Taking it to the next level, networking also informed her worldview and changed her personal identity. The lasting imprint of Eastern culture on the Western imagination was something that connected Graham to many other Western travelers and temporary (or permanent) migrants to the East, including Robert Morrison (per Hargrave's essay), Lord George Macartney (per Clingham's essay), and to some extent even to Aubin's fictional characters (per Cahill's essay). It also linked her to those who stayed home but "participated vicariously" (as Richard Coulton puts it) in the experience of the East, made accessible through commodities, ideas, paintings, and travel writing, all of which circulated back and forth through intersecting networks thematized in the essays by Coulton, Howard-Smith, Watt, and me.

Born into a family of professional world citizens and brought up as a cosmopolitan, Maria Graham was well prepared for adult life as a global nomad whose accumulated output of travel narratives could, in the amalgam, be read as a networked autobiography. The center of her globalized universe, Cope argues, was to be found in the East rather than the West despite her British roots. Graham's transculturated biography enabled her to look at the Far West (of Chile and Brazil) not through the Eurocentric lens of her nominal home country (which informed the travel writing of many of her contemporaries) but rather through the lens of the Far East (especially India) that she had visited earlier in life and whose indelible imprint on her imagination transformed it into a permanent and omnipresent experiential matrix. The latter appears like a radicalized version of the orientalization experienced by retired nabobs-turned-gardeners. That Graham managed six ocean crossings with extended stays in five countries by her mid-thirties and that, wherever she went, she received visitors from every corner of assorted empires, demonstrates the importance of improved, faster travel technology for the maintenance of cultural networks, which require the quick and continuous exchange of information and which evoke an imagined sense of propinquity: a sense that cultures are readily accessible, connected, and not so different as to interfere with compatibility.

Written by scholars whose expertise lies in Western languages, literatures, and cultures, *Oriental Networks* explores East-West connectivity during the long eighteenth century from Western perspectives. By tracing extant and emerging forms of contact that facilitated intercultural exchange and by tracing Western individuals and groups that engaged with the East and found themselves transculturated in the process, the volume might encourage further network-centered research that would approach the subject from the opposite direction.[54] Specialists with expertise in Eastern languages, literatures, and cultures have already begun to rediscover additional missing links of a historical web of collaborating individuals and groups in East and West[55] that can be reconstructed only by more interdisciplinary *scholarly* networking.

Notes

1. See, for instance, Stanley Wasserman and Katherine Faust, *Social Network Analysis: Methods and Applications* (Cambridge: Cambridge University Press, 1994), and Linton C. Freeman, *The Development of Social Network Analysis: A Study in the Sociology of Science* (Vancouver, BC: Empirical, 2004).

2. See, for instance, David Blakesley and Thomas Rickert, "From Nodes to Nets: Our Emerging Culture of Complex Interactive Networks," *JAC: A Journal of Composition Theory* 24, no. 4 (2004), 821–830.

3. Albert-László Barabási, *Linked: The New Science of Networks* (Cambridge, MA: Perseus, 2002); *Bursts: The Hidden Patterns behind Everything We Do* (New York: Dutton, 2010); for more specialized audiences, also Albert-László Barabási and Réka Albert, "Emergence of Scaling in Random Physics Networks," *Science* 286 (2013), 509–512.

4. Subtitle of a 2014 paperback edition of Barabási, *Linked*.

5. Bruno Latour, *Les microbes: Guerre et paix, suivi de irréductions* (Paris: Editions A. M. Métailié, 1984), trans. Alan Sheridan and John Law as *The Pasteurization of France* (Cambridge, MA: Harvard University Press, 1988); *Aramis, ou, L'amour des techniques* (Paris: La Découverte, 1992), trans. Catherine Porter as *Aramis, or, The Love of Technology* (Cambridge, MA: Harvard University Press, 1996). For networks in social theory see also Bruno Latour, *Reassembling the Social: An Introduction to Actor-Network-Theory* (Oxford: Oxford University Press, 2005) and Manuel Castells, *The Information Age: Economy, Society and Culture*, vol. 1, *The Rise of the Network Society* (Cambridge, MA: Blackwell, 1996).

6. For Barabási, see, for instance, Daniel B. Stouffer, R. Dean Malmgren, and Luís A. Nunes Amaral, "Comment on *The Origin of Bursts and*

Heavy Tails in Human Dynamics," October 25, 2005. https://arxiv.org
/pdf/physics/0510216.pdf; Walter Willinger, David Alderson, and
John C. Doyle, "Mathematics and the Internet: A Source of Enormous
Confusion and Great Potential," *Notices of the American Mathematical
Society* 56, no. 5 (2009), 586–599; Lior Pachter and Nicolas Bray, "The
Network Nonsense of Albert-László Barabási," *Bits of DNA*, Febru-
ary 10, 2014, blog, https://liorpachter.wordpress.com/2014/02/10/the
-network-nonsense-of-albert-laszlo-barabasi/. Critical responses to
Latour's work include Harry M. Collins and Steven Yearley, "Episte-
mological Chicken," in *Science as Practice and Culture*, ed. Andrew
Pickering (Chicago: University of Chicago Press, 1992), 301–326;
Langdon Winner, "Upon Opening the Black Box and Finding it
Empty: Social Constructivism and the Philosophy of Technology,"
Science, Technology, & Human Values 18, no. 3 (1993), 362–378; and
Andrea Whittle and André Spicer, "Is Actor Network Theory
Critique?" *Organization Studies* 29, no. 4 (2008), 611–629.

7. This was the title of an interdisciplinary lecture series held at the
University of Illinois, Urbana-Champaign in 2006: http://www.ncsa
.illinois.edu/Conferences/Networks/.

8. See, for instance, Alan Lester, "Imperial Circuits and Networks:
Geographies of the British Empire," *History Compass* 4, no. 1 (2006),
131; for Latour, also Whittle and Spicer, "Is Actor Network Theory
Critique?," 9–10.

9. See, for instance, Alexander R. Galloway and Eugene Thacker, *The
Exploit: A Theory of Networks* (Minneapolis: University of Minnesota
Press, 2007), 13, who state that "the mere existence of this multiplicity
of nodes in no way implies an inherently democratic, ecumenical, or
egalitarian order" and that "there exist new modes of control entirely
native to networks, modes that are at once highly centralized and
dispersed, corporate and subversive."

10. David Cohen, "All the World's a Net," *New Scientist*, 174, no. 2338
(April 13, 2002), 24–29.

11. See Cohen, "All the World's a Net," 28.

12. [Joseph Addison], *Spectator* 69 (May 19, 1711), in Joseph Addison and
Richard Steele, *The Spectator*, ed. Donald F. Bond (Oxford: Clarendon,
1965), 1:296.

13. Oliver Goldsmith, "Letter XX," from *The Citizen of the World: Or Letters from a Chinese Philosopher Residing in London to His Friends in the East*, 2 vols. (1762), in *Collected Works of Oliver Goldsmith*, ed. Arthur Friedman (Oxford: Clarendon, 1966), 2:86.

14. John Dyer, *The Fleece. A Poem. In Four Books.* (London: R. and J. Dodsley, 1757), bk. 4, ll. 673–676.

15. Addison, Spectator 69 (May 19, 1711), in *The Spectator*, ed. Bond, 1:294–295.

16. Goldsmith, "Letter XI," from *Citizen of the World*, in *Collected Works*, 2:52.

17. William Wordsworth, "Composed Upon Westminster Bridge, September 3, 1802," in *Complete Poetical Works with Introductions and Notes*, ed. Thomas Hutchinson; rev. ed. Ernest de Selincourt (Oxford: Oxford University Press, 1988), 214.

18. The danger for world peace emanating from modern decentralized power structures such as transnational networks of terrorism and counterterrorism (discussed in detail in Galloway and Thacker's "Prolegomenon," *Exploit*, 1–22), on the other hand, seems to be a topic without eighteenth-century precedents.

19. Sankar Muthu, "Adam Smith's Critique of International Trading Companies: Theorizing 'Globalization' in the Age of Enlightenment," *Political Theory* 36, no. 2 (2008), 185–212.

20. Adam Smith, *An Inquiry into the Nature and Causes of the Wealth of Nations* (1776), ed. R. H. Campbell, A. S. Skinner, and W. B. Todd (Oxford: Oxford University Press, 1976), 2:626.

21. Smith, *Wealth of Nations*, 2:661.

22. See, for instance, Alan Lester, *Imperial Networks: Creating Identities in Nineteenth-Century South-Africa and Britain* (London: Routledge, 2001), and "Imperial Circuits"; Juliette Bridgette Milner-Thornton, "Imperial Networks in a Translational Context," in *The Long Shadow of the British Empire: The Ongoing Legacies of Race and Class in Zambia* (New York: Palgrave Macmillan, 2012), 107–129; Barry Crosbie, *Irish Imperial Networks: Migration, Social Communication and Exchange in Nineteenth-Century India* (Cambridge: Cambridge University Press, 2012); and John M. Mackenzie, "Imperial Networks," *Britain and the World* 11, no. 2 (2018), 149–152.

23. Anna Letitia Barbauld, "Epistle to William Wilberforce, Esq. On the Rejection of the Bill for Abolishing the Slave Trade," in *Selected Poetry*

and Prose, ed. William McCarthy and Elizabeth Kraft (Peterborough, ON: Broadview, 2002), 126, ll. 96–99. I thank James Watt for drawing my attention to this aspect of Barbauld's poem.

24. Galloway and Thacker, *Exploit*, 32–34, contest earlier definitions of networks as insufficient and emphasize additional characteristics such as a high degree of "internal complexity" and heterogeneity.

25. Smith, *Wealth of Nations*, 1:448, my italics.

26. James Beattie, Edward Melillo, and Emily O'Gorman, "Rethinking the British Empire through Eco-Cultural Networks: Materialist-Cultural Environmental History, Relational Connections and Agency," *Environment and History* 20, no. 4 (2014), 561–575.

27. See, for instance, Eric Darnton, *Poetry and the Police: Communication Networks in Eighteenth-Century Paris* (Cambridge, MA: Belknap Press of Harvard University Press, 2010); Hilary Brown and Gillian Dow, eds., *Readers, Writers, Salonnières: Female Networks in Europe, 1700–1900* (New York: Peter Lang, 2011); Tijl Vanneste, *Global Trade and Commercial Networks: Eighteenth-Century Diamond Merchants* (London: Pickering and Chatto, 2011); Andreas Gestrich and Margrit Schulte Beerbühl, eds., *Cosmopolitan Networks in Commerce and Society, 1660–1914* (London: German Historical Institute, 2011); Carson Bergstrom, "Literary Coteries, Network Theory, and the Literary and Philosophical Society of Manchester," *ANQ: A Quarterly Journal of Short Articles, Notes, and Reviews* 26, no. 3 (2013), 180–188; Diane E. Wenger, *A Country Storekeeper in Pennsylvania: Creating Economic Networks in Early America, 1790–1807* (University Park: Pennsylvania State University Press, 2013); Ileana Baird, ed., *Social Networks in the Long Eighteenth Century: Clubs, Literary Salons, Textual Coteries*, ed. (Newcastle upon Tyne: Cambridge Scholars, 2014); Albrecht Classen, "A Global Epistolary Network: Eighteenth-Century Jesuit Missionaries Writing Home with an Emphasis on Philipp Segesser's Correspondence from Sonora/Mexico," *Studia Neophilologica* 86 (2014), 79–94; Lindsay O'Neill, *The Opened Letter: Networking in the Early Modern British World* (Philadelphia: University of Pennsylvania Press, 2014); Jon Mee, ed., "Networks of Improvement: Literary Clubs and Societies c. 1760–c. 1840," special issue, *Journal for Eighteenth-Century Studies* 38, no. 4 (2015), 475–612; Mark C. Wallace and Jane Rendall, eds.,

Association and Enlightenment: Scottish Clubs and Societies, 1700–1830 (Lewisburg, PA: Bucknell University Press, 2020).

28. See, for instance, Felicity A. Nussbaum, ed., *The Global Eighteenth Century* (Baltimore, MD: Johns Hopkins University Press, 2003); Humberto Garcia, *Islam and the English Enlightenment, 1670–1840* (Baltimore, MD: Johns Hopkins University Press, 2011); Michael Rotenberg-Schwartz and Tara Czechowski, eds., *Global Economies, Cultural Currencies of the Eighteenth Century* (New York: AMS, 2012); Ileana Baird and Christina Ionescu, eds., *Eighteenth-Century Thing Theory in a Global Context: From Consumerism to Celebrity Culture* (Farnham: Ashgate; 2013); Evan Gottlieb, *Romantic Globalism: British Literature and Modern World Order, 1750–1830* (Columbus: Ohio State University Press, 2014); Kevin L. Cope and Samara Anne Cahill, eds., *Citizens of the World: Adapting in the Eighteenth Century* (Lewisburg, PA: Bucknell University Press, 2015); Nabil Matar, "The Arabic-Speaking Peoples and 'Globalization': The Eighteenth Century," *Eighteenth Century: Theory and Interpretation* 58, no. 1 (2017), 121–125.

29. See, for instance, David Porter, *Ideographia: The Chinese Cypher in Early Modern Europe* (Stanford, CA: Stanford University Press, 2001); Markman Ellis, *The Coffee House: A Cultural History* (London: Weidenfeld and Nicolson, 2004); Robert Markley, ed., "Europe and East Asia in the Eighteenth Century," special issue, *Eighteenth Century: Theory & Interpretation*, 45, no. 2 (2004), 111–206; Robert Markley, "Gulliver and the Japanese: The Limits of the Postcolonial Past," *Modern Language Quarterly: A Journal of Literary History* 65, no. 3 (2004), 457–479; Eugenia Zuroski, "Disenchanting China: Orientalism and the Aesthetics of Reason in the English Novel," *Novel: A Forum on Fiction* 38, no. 2/3 (2005), 254–271; Markman Ellis, ed., *Eighteenth-Century Coffee-House Culture*, 4 vols. (London: Pickering and Chatto, 2006); Robert Markley, *The Far East and the English Imagination, 1600–1730* (Cambridge: Cambridge University Press, 2006); Ellis, Markman, Richard Coulton, Matthew Mauger, et al., eds., *Tea and the Tea-Table in Eighteenth-Century England*, 4 vols. (London: Pickering and Chatto, 2010); Robert Markley, "China and the Making of Global Modernity," *Eighteenth-Century Studies* 43, no. 3 (2010), 299–339; Debra Johanyak and Walter S. H. Lim, eds., *The English Renaissance,*

Orientalism, and the Idea of Asia (New York: Palgrave Macmillan, 2010); Barbara Schmidt-Haberkamp *Europa und die Türkei im 18. Jahrhundert/Europe and Turkey in the 18th Century*, ed. (Bonn: Bonn University Press; Göttingen: V. & R. Unipress, 2011); Chi-ming Yang, *Performing China: Virtue, Commerce, and Orientalism in Eighteenth-Century England, 1660–1760* (Baltimore, MD: Johns Hopkins University, 2011); Peter J. Kitson, *Forging Romantic China: Sino-British Cultural Exchange, 1760–1840* (Cambridge: Cambridge University Press, 2013); James Mulholland, *Sounding Imperial: Poetic Voice and the Politics of Empire, 1730–1820* (Baltimore: Johns Hopkins University Press, 2013); James Mulholland, "Connecting Eighteenth-Century India: Orientalism, Della Cruscanism, and the Translocal Poetics of William and Anna Maria Jones," in *Representing Place in British Literature and Culture, 1660–1830: From Local to Global*, ed. Evan Gottlieb and Juliet Shields (Farnham: Ashgate, 2013), 117–136; Dermot Ryan, *Technologies of Empire: Writing, Imagination, and the Making of Imperial Networks* (Newark: University of Delaware Press, 2013); Peter M. Solar, "Opening to the East: Shipping between Europe and Asia, 1770–1830," *Journal of Economic History* 73, no. 3 (2013), 625–661; Eugenia Zuroski Jenkins, *A Taste for China: English Subjectivity and the Prehistory of Orientalism* (Oxford: Oxford University Press, 2013); Simon Davies, Daniel Sanjiv Roberts, and Gabriel Sánchez Espinosa, *India and Europe in the Global Eighteenth Century* (Oxford: Voltaire Foundation, 2014); Robert Markley, "China and the English Enlightenment: Literature, Aesthetics, and Commerce," *Literature Compass* 11, no. 8 (2014), 517–527; Ernest W. B. Hess-Lüttich and Yoshito Takahashi, eds., *Orient im Okzident—Okzident im Orient: West-Östliche Begegnungen in Sprache und Kultur, Literatur und Wissenschaft* (Frankfurt: Peter Lang, 2015); Evan Gottlieb, ed., *Global Romanticism: Origins, Orientations, and Engagements, 1760–1820* (Lewisburg, PA: Bucknell University Press, 2015); Markman Ellis, Richard Coulton, and Matthew Mauger, *Empire of Tea: The Asian Leaf that Conquered the World* (London: Reaktion, 2015); Greg Clingham, "Cultural Difference in George Macartney's *An Embassy to China, 1792–94*," *Eighteenth-Century Life* 39, no. 2 (2015), 1–29; Peter J. Kitson and Robert Markley, eds., *Writing China: Essays on the Amherst Embassy (1816) and Sino-British*

Cultural Relations (Cambridge: D. S. Brewer, 2016); Pim de Zwart, "Globalization in the Early Modern Era: New Evidence from the Dutch-Asiatic Trade, c. 1600–1800," *Journal of Economic History* 76, no. 2 (2016), 520–558; William Farrell, "Smuggling Silks into Eighteenth-Century Britain: Geography, Perpetrators, and Consumers," *Journal of British Studies* 55, no. 2 (2016), 268–294; Mario Cams, "Not Just a Jesuit Atlas of China: Qing Imperial Cartography and Its European Connections," *Imago Mundi* 69, no. 2 (2017), 188–201.

30. For notable inroads in the field with a special focus on the nineteenth and early twentieth centuries, see James Beattie, Edward Melillo, and Emily O'Gorman, eds., *Eco-Cultural Networks and the British Empire: New Views on Environmental History* (London: Bloomsbury, 2015), especially the contributions by Joseph Lawson, Edward D. Melillo, Eugenia W. Herbert, James Beattie, and Robert Peckham; with a regional focus also Su Fang Ng, ed., "Transcultural Networks in the Indian Ocean, Sixteenth to Eighteenth Centuries: Europeans and Indian Ocean Societies in Interaction," special issue, *Forms of Discourse and Culture* 48, no. 2 (2015), 119–340, especially the essays by Timothy Davies, Robert Markley, and Amrita Sen.

31. James Beattie, Edward Melillo, and Emily O'Gorman, "Rethinking the British Empire through Eco-Cultural Networks: Materialist-Cultural Environmental History, Relational Connections and Agency," *Environment and History* 20, no. 4 (2014), 571.

32. Mark Casson, "Networks in Economic and Business History: A Theoretical Perspective," in Gestrich and Schulte Beerbühl, *Cosmopolitan Networks*, 49.

33. Beattie, Melillo, and O'Gorman, "Rethinking the British Empire," 571.

34. See, for instance, Edward Said, *Orientalism* (New York: Pantheon, 1978).

35. This observation resonates with arguments by academic orientalists who have pointed out the demonstrable appreciation of Eastern culture evident in the work of some Western eighteenth- and nineteenth-century pioneers of Asian studies. See, for instance, O. P. Kejariwal, *The Asiatic Society of Bengal and the Discovery of India's Past*, (Delhi: Oxford University Press, 1988), and Robert Irwin, *Dangerous Knowledge: Orientalism and Its Discontents* (Woodstock and New York: Overlook, 2006).

36. Daniel Goffman, *The Ottoman Empire and Early Modern Europe* (Cambridge: Cambridge University Press, 2002), 6.

37. Julie F. Codell, "The Art of Transculturation," in *Transculturation in British Art, 1770–1930*, ed. Julie F. Codell (Farnham: Ashgate, 2012), 1.

38. Fernando Ortiz, *Cuban Counterpoint: Tobacco and Sugar*, trans. Harriet de Onís (New York: Alfred A. Knopf, 1947), 98.

39. Ortiz, *Cuban Counterpoint*, 102–103.

40. Wolfgang Welsch, "Transculturality: The Puzzling Form of Cultures Today," in *Spaces of Cultures: City—Nation—World*, ed. Mike Featherstone and Scott Lash (London: Sage, 1999), 196 and 205. I thank Stephanie Howard-Smith for bringing to my attention more recent appreciations of Ortiz by Codell, Welsch, Cooper and White, as well as Mark Millington's "Transculturation: Taking Stock," in *Transculturation: Cities, Spaces and Architectures in Latin America*, ed. Felipe Hernández, Mark Millington, and Iain Borden (Amsterdam: Rodopi, 2005), 204–234.

41. Codell, "Art of Transculturation," 4.

42. Rose Cooper and Darcy White, "Teaching Transculturation: Pedagogical Processes," *Journal of Design History* 18, no. 3 (2005), 286.

43. I thank Richard Coulton for contributing to this paragraph through an informal written exchange.

44. Cooper and White, "Teaching Transculturation," 286.

45. Lester, "Imperial Circuits," 124.

46. Lester, "Imperial Circuits," 131.

47. On the fragility of empires, which are "not just structures but processes as well," and, as such, "constantly remade and reconfigured," see Tony Ballantyne, *Orientalism and Race: Aryanism in the British Empire* (Basingstoke: Palgrave, 2002), 39.

48. See also Millington, "Transculturation," 223, who emphasizes its ability to capture "a sense of the plurality of movement between cultures."

49. Ortiz, *Cuban Counterpoint*, 103.

50. For interactions of individual elements in complex networks as adaptive systems, see also Blakesley and Rickert, "From Nodes to Nets," 823.

51. I thank Jennifer Hargrave for allowing me to use excerpts from an informal personal exchange for my contextualization of her research.

52. Gestrich and Schulte Beerbühl, *Cosmopolitan Networks*, 8–9.

53. I am indebted to James Watt for a private written exchange from which I quote in this paragraph.

54. For pioneering studies with a reverse perspective, see, for instance, KumKum Chatterjee and Clement Hawes, eds., *Europe Observed: Multiple Gazes in Early Modern Encounters* (Lewisburg, PA: Bucknell University Press, 2008); Nabil Matar, *Europe Through Arab Eyes, 1578–1727* (New York: Columbia University Press, 2009); and Gerald MacLean and Nabil Matar, *Britain and the Islamic World, 1558–1713* (Oxford: Oxford University Press, 2011). I thank Greg Clingham for drawing my attention to these works and for his constructive criticism of my introduction to this volume.

55. See, for instance, Chi-ming Yang, ed., "Eighteenth-Century Easts and Wests," special issue, *Eighteenth-Century Studies* 47, no. 2 (2014), 95–231.

1

Knowing and Growing Tea

China, Britain, and the Formation of a Modern Global Commodity

RICHARD COULTON

The consumption of tea is widely observed to be a marker of British national identity, both within and beyond the United Kingdom. This signification for the habit of imbibing the infusion of a desiccated foreign leaf was forged during the eighteenth century, a period when (almost) all tea in Britain was sourced from China.[1] An exotic luxury in Restoration London, where it was favored by genteel savants and fashionable ladies at court, tea was considered within the means of all "well-regulated Families" (that is to say, of the upper-middling sort) in the metropolis by 1711, and was judged in evidence presented before the Commissioners of Excise to have "found an entrance into every Cottage" in the kingdom by the late 1770s.[2] That taking tea became a universal British practice was by no means inevitable, any more than it was the calculated outcome of a single strategic plan. Rather the interaction of multiple and complex processes—operating across networks of commerce, society, culture, and knowledge—gradually generated conditions within which the nationwide distribution of an unprecedented article of grocery became possible, profitable, and sustainable. Each of these processes relied variously upon forms of exchange with

East Asia. To drink tea in eighteenth-century Britain was to participate in rituals of being and traditions of knowing that traversed geographical, political, and historical boundaries. When tea chests were landed wholesale at London's Legal Quays, Britain imported not just the core ingredient for a hot drink but also aspects of its Chinese heritage, pertaining to practices of taste, sociability, understanding, and well-being.

So remarkable and so pervasive was tea's adoption within the British context that it can be claimed as the first modern global commodity. There are three elements to this contention. First, tea was—and today still is—produced and consumed cheaply by laborers on opposite sides of the planet, while vastly enriching the owners of intermediary supply chains. During the eighteenth century, moreover, tea was no product of colonial agriculture but rather an article of mercantile exchange between nations geographically and politically remote from one another. Second, tea has no immediate value for life—any nutritional benefits derive from the water that it flavors, or the milk and sugar with which it may be mixed—yet in Britain and elsewhere it has achieved the status of a social and economic necessity. In this way tea epitomizes the modern commodity, a nonessential item of commerce that nonetheless becomes integral to ordinary people's daily practices. Third, although the incomes and living standards of Britons remained generally stagnant throughout the eighteenth century, tea reached (and has since retained) a mass market, heralding the advent of a consumer-oriented society.[3] Tea's saturation of the British marketplace was unprecedented for a globally sourced article of trade that no one needed and yet everyone seemed to want.

The first part of this essay examines aspects of tea's formation as "modern global commodity," by sketching the broad trajectory of its initial transculturation between China and Britain.[4] In the second part the focus narrows to examine attempts by European bioprospectors during the mid-century to transplant and naturalize the tea shrub as indigenous—or, alternatively, colonial—cash crop.[5] Added to its intrinsic (if mild) psychotropic properties, the social and cultural dimensions of tea's accommodation to British consumers configure it as a vibrant case study for interrogating the types of "oriental network" with which this volume is preoccupied.

Tea and the Formation of a Modern Global Commodity

Tea has been consumed in China for millennia. The eighth-century sage Lu Yu teaches in his *Chajing* or "Classic of Tea" (760)—the most highly regarded of all early Chinese writings on the beverage— that "tea, used as a drink, was first discovered by the Emperor Shen Nung" (Shennong), a legendary ruler of China whose reign is customarily dated 2737–2698 B.C.E.[6] There is certainly unequivocal evidence for its preparation and consumption by the first century B.C.E.[7] By the time that European merchants and missionaries began arriving concertedly in East Asia some 1500 years later, tea had become rooted immemorially within Chinese commerce and culture. Feted as a medical panacea, it was also focal within religious, political, and social ceremony. In his "Treatise on the Things of China" (*Tratado das Cousas da China*), the Portuguese Dominican friar Gaspar da Cruz (1520–1570) describes regular encounters with tea on the island of Lampacau in the Pearl River (Zhujiang) Delta. "Whatsoeuer person or persons come to any mans house of qualitie," da Cruz recalls, "hee hath a custome to offer him in a fine basket one Porcelane [cup] . . . with a kinde of drinke which they call *Cha* [tea], which is somewhat bitter, red, and medicinall."[8] Half a century later Alexandre de Rhodes (1591–1660), a French Jesuit who spent the 1630s on Macau, documented his personal experience of the infusion as both a cure for headaches and an aid to wakefulness when fulfilling priestly duties at night-time.[9] Meanwhile Johan Nieuhof (1618–1672), the steward to a Dutch commercial embassy visiting the court of the Shunzhi Emperor in 1656, observed the ritual consumption of the hot infusion in that loftiest of settings:

> The Emperour . . . sat about thirty paces from the Embassadours, his Throne so glitter'd with Gold and Precious Stones,

that the Eyes of all that drew near dazzled: . . . next and on his side sat the Vice-Roys, Princes of the Blood, and all other great Officers of the Court, all likewise drinking *Thea* in Wooden Dishes, and that in great abundance.[10]

These descriptive snatches stirred awareness of and interest in tea among merchants and natural philosophers in early modern Europe. Few immediately grasped its commercial potential—de Rhodes was perhaps the first seriously to do so—but travelers to China did repeatedly remark upon tea's social prevalence, its alleged medicinal properties, its visibility within the staging of encounters between high-status officials, its capacity to induce both alertness and tranquility in religious adherents. By the second quarter of the seventeenth century, small quantities of leaf tea were being imported through Amsterdam by the Dutch East India Company, at first from Japan but later (as demand grew) also from China.[11] It was probably via Holland that the earliest shipments reached London, where on September 23, 1658, the businessman Thomas Garway (d. 1692?) advertised in the news periodical *Mercurius Politicus* that the "Excellent, and by all Physitians approved, *China* Drink, called by the *Chineans*, *Tcha*, by other Nations *Tay alias Tee*, is sold at the *Sultaness-head*, a *Cophee-house* in *Sweetings* Rents by the Royal Exchange."[12] The following year Thomas Rugg, a barber in Covent Garden, noted in his diary that there was "att this time a Turkish drink to bee sould, almost in evry street, called Coffee, and another kind of drink called Tee, and also a drink called Chacolate, which was a very harty drink."[13] Alongside two comparable novelties, tea was discovering a receptive market. Together these drinks represented a departure for English consumers. They were taken hot and tasted bitter; they proved expensive to purchase yet became increasingly fashionable; they were the global products of distant continents; their reception in England had been extensively mediated by travelers' tracts and medical writing during preceding decades. Of the three, it was tea that would most effectively colonize Britons' palates during the century to come.

Thomas Garway, keen to emphasize his cachet as the earliest retailer of tea in London, understood that one strategic key to unlocking demand for this innovative beverage was to educate consumers about its East Asian origins. During the 1660s he published at least two versions of a marketing broadside titled *An Exact Description of the Growth, Quality and Vertues of the Leaf Tea* that detailed the commodity's botanical features and agricultural cultivation, its artisanal preparation and social appeal, its pharmaceutical virtues and physiological benefits. In doing so, Garway explicitly and repeatedly cites European accounts of tea in China, at once reassuring patrons of its wholesomeness, while tantalizing them with the promise of participating vicariously in that exotic realm through the activity of domestic consumption. Yet for all his professed tutelage in tea preparation by "the most knowing Merchants and Travellers into those Eastern Countries," Garway was also crucially transforming tea for its new context. His register of medical preparations recommends the addition of milk, while his shop in Exchange Alley variously dispenses *"Tea, Coffee* and *Chocolat."* This series of combinations was unique to the drink's early modern European situation.[14]

At sixteen to fifty shillings per pound, the tea that Garway sold was well beyond the pockets of all but the "Nobleman, Physitians, Merchants and Gentlemen of Quality" whom his flyer overtly addressed. In the years following the Restoration, tea remained sufficiently valuable and curious for the East India Company to include a modest package among rarities presented as a commercial tribute to Charles II.[15] But while the fashion for the beverage erratically peaked and troughed among well-heeled metropolitan consumers during the decades that followed, from the earliest years of the eighteenth century the supply of tea from China stabilized and then consistently grew. Annual tea imports to Britain were below 10,000 pounds for the last time in 1705 and have exceeded 100,000 pounds since 1710; they surpassed one million pounds for the first time in 1723 and have dipped beneath that mark on only three occasions since (the last time in 1747). Given that one government estimate from the period (albeit several decades later) calculated the yearly consumption of "a Common Family" at ten pounds, this means that up to 100,000 households nationwide were regularly purchasing tea by the second quarter of the century.[16] This relentless growth and a related shift in habits and attitudes were noted by contemporaries. The physician Thomas Short had considered the "Infusion" to be "universal" in 1730 (although he did except "Persons . . . of the very lowest Rank"); while the Methodist preacher John Wesley's ad hoc interrogation of the liquid diet of "the Abundance of People in *London*" during the mid-1740s was apparently met with the response that "I drink scarce any Thing but a little Tea, Morning and Night."[17] Interestingly when considering the longstanding British preference for darker oxidized leaves, almost all tea imported from China prior to 1700 was green, as opposed to black (or, more accurately, oolong) varieties. Green teas retained a 50 percent market share until 1750 and remained at 20 percent as late as the 1830s, being absolutely supplanted only in the later nineteenth century when the United Kingdom switched its attention to managing production in colonial India.[18]

As tea suffused the daily practices of Britons, so it also permeated the realm of culture. As a means for signaling taste and wealth, tea suggested itself as a focus for painters and poets. Nahum Tate's long poem *Panacea* (1700) offers two legends of origin for tea, narrated by a Wiltshire shepherd bearing the unlikely name Palaemon. In Canto I, tea is gifted by Confucius to China, "a timely Cure" for the "Publick Grief" that had attended the depredations of a corrupt monarch, Xia Jie (1728–1675 B.C.E.). In Canto II, the classical gods of Olympus compete for the title of tea's patron, aiming variously to appropriate its apparent capacity to bestow qualities such as health, wisdom, and chastity. The dispute is resolved only when Jove elevates tea—or *Thea*—herself to the pantheon: now "*GODDESS* was its *Name*," runs the poet laureate's closing pun. Tate's innovative if sometimes ungainly fusion of Asian and European mythology demonstrates how tea imported elements of its Chinese heritage, while also being reconfigured for its transformed context.[19] Correspondingly, in Richard Collins's group portrait *The Tea Party* (c. 1727), the material life of China is

FIGURE 1.1. Richard Collins, *The Tea Party*, circa 1727, oil on canvas, 100×120 cm, Goldsmiths' Hall, London. © The Goldsmiths' Company.

integrated within the quotidian environment of the sitters: they wear silk, they touch porcelain, they ingest tea (Figure 1.1). But here, once more, the beverage is made at home within a new situation. The leisured and affectionate domesticity of a genteel English family at tea is arranged by Collins to become the subject of a conversation piece painting, one no doubt commissioned to articulate conspicuously on-trend affluence.

Tea's domains of signification in cultural production were complex and contested. As a sober yet social beverage, at once calming and reviving for the drinker, tea came to symbolize polite—and, above all, female—sociability. "I shall take it for the greatest Glory of my Work," announces the Spectator as early as his fourth number, "if among reasonable Women this Paper may furnish *Tea-table Talk*."[20] Yet by the same token, the generation of this uncertainly regulated and socially accessible space (the "Tea-table") unsettled established homosocial conventions of alehouse and tavern drinking, while provocatively implying a potential to nurture scandal and gossip. In *Moll Flanders* (1722), the narrator assists a young widow of Redriff (on the south bank of the Thames between London and Greenwich) to come to favorable financial terms with the rakish ship's captain who woos her. Moll and her friend deploy the power of calculated rumor to unsettle his reputation and self-confidence: "for telling her Story in general to a Couple of Gossips in the Neighbourhood, it was the Chat of the Tea Table all over that part of the Town, and I met with it wherever I visited."[21] Moreover, as a luxurious Asian commodity, tea may have been a harbinger of cosmopolitanism and commercial prosperity, but it was also suspiciously foreign, a usurper of indigenous goods that bled sterling silver into Chinese coffers. In William Hogarth's *Portrait of a Family* (c. 1735), the central couple drink tea from a delicate equipage, framed by a large porcelain vase (in the fireplace) and imposing lacquerware cabinet (in the corner), all sumptuous fruits of the trade with China (Figure 1.2). As in Collins's *The Tea Party*, tea is figured as a domestic occasion at ease with the opulence invigorated by Britain's global reach. Yet beneath the surface are hints of anxiety about the basis and security of this family's comfort. An excited kitten overturns the yarn basket that supplies the tapestry table, while several sitters appear disquieted by the intrusion of a servant—probably an African boy—who carries a second tea-tray bearing an upturned cup (this figure was apparently excised when the left of the canvas was cropped, sparing only his right arm and the skirt of his livery coat).

During the decades around the turn of the eighteenth century, a discourse of tea was consolidated in Britain, anatomizing and reimagining its botany, its cultivation, its procurement, its consumption, its propensities, its symbolism. The British consumer-subject's desire (and ultimately need) for tea was generated through the nexus of social processes and global exchange from which this discourse emerged and which in turn it reinforced. To assign a disciplinary power to tea might be to overstate the Foucauldian

FIGURE 1.2. William Hogarth, *Portrait of a Family*, circa 1735, oil on canvas, 53.3×74.9 cm, Yale Center for British Art, Paul Mellon Collection. Courtesy of the Yale Center for British Art under the terms of Yale University's Open Access Policy.

inflections of this claim, but tea subtly insinuates itself into bodies, minds, habits.[22] It changes behaviors; it connotes its own province in space (the tea table) and society (the tea party); it determines its own peculiar temporal moment (teatime). As tea's social incidence and material visibility increased in Britain, so it transitioned from the precinct of the remarkable to that of the everyday, transporting with it this ideological cargo. If not yet wholly naturalized, tea was decidedly transculturated. In other words, it retained traces of some of its meanings and operation within China while losing others as it negotiated new purposes for itself in Europe. Tea's formation as a modern global commodity that reached a mass market in Britain intrinsically depended upon this production of a discourse of tea. It is to one dimension of this discourse, the botanical understanding of Chinese tea within European systems of knowledge, that the second part of this essay now turns.

At the outset of this essay, I suggested that when Britain imported tea, the commodity carried with it aspects of its Chinese heritage. This was not merely a gradual diffusion but an active undertaking. As the example of Garway's broadside advertisement illustrates, the production of knowledge about tea in Britain was crucial to its widespread adoption and growing popularity. Historians of science have argued convincingly in recent decades that trading institutions like the East India Company invested both directly and indirectly in the natural historical researches of their servants in far-flung territories. They understood that maximizing commercial returns relied not just upon abstract mercantile nous, but also upon learning about, adapting to, and exploiting the new environments—natural and cultural—within which their employees operated.[23] This was vital both for successfully bringing to market new goods such as drugs, foodstuffs, and dyes, and for ensuring the sustainable alimentary and pharmacological supply of ships and factories within unfamiliar and often inhospitable terrains. By the same token, the most prestigious and effective natural history collectors in Europe overwhelmingly depended for botanical and zoological specimens on the networks of educated surgeons, clergymen, and merchants who traveled with commercial enterprises to distant zones such as the Indian Ocean world.[24] The interests of commercial expansion and natural knowledge frequently comingled and underwrote one another within the multiple theaters of Britain's growing global ambitions. The case of the emerging trade in tea was no exception.

Throughout the seventeenth century, writings by travelers to East Asia had offered snippets of tea lore to the curious, but there had been no concerted description of the plant's botany, cultivation, or manufacture for consumption. This changed around the turn of the eighteenth century when two East India Company servants issued accounts of tea derived from their own experience. The first was the clergyman John Ovington (c. 1653–1731), who served as the salaried chaplain to the factory at Surat in northwest

India between 1689 and 1693. Following his return to England, Ovington composed *A Voyage to Suratt* (1696), a lengthy octavo work that tracks the itinerary of his journey to and from South Asia, variously detailing items of geographical, political, natural historical, anthropological, and commercial interest. Ovington encountered Surat as a cosmopolitan entrepôt. Among those with whom he mixed were diplomatic and merchant visitors from China, and "*Bannian*" tradesmen from Gujarat (Hindu locals whom he distinguished from the city's Islamic "*Moor* Inhabitants"). Among the Bannians in Surat, Ovington noticed, the use of tea was particularly prevalent (although it was also the "common Drink" of European settlers), while from the transient Chinese community he recorded knowledge about the complex preparation and multiple varieties—"Bing, Singlo, Bohe"—of dried tea-leaves.[25] This specific typology of tea as consumer product was hitherto virtually unknown in England, but rapidly acquired common currency in the decade that followed.

Presumably encouraged by the reception of his *Voyage*, which earned him a modest cash perquisite from the directors of the East India Company, and by the drink's evident cachet in England, Ovington next drew up *An Essay upon the Nature and Qualities of Tea* (1699).[26] This treatise augmented his immediate knowledge of the beverage from Surat with opinions drawn from scholarly authorities concerning tea's botanical properties and pharmaceutical virtues, as well as observations pertaining to the wholesale tea trade in China (a "large and flourishing Empire") and advice directing Europeans how to dodge the "Cheats and Frauds" of local merchants in order "to trade therein with Advantage."[27] Such guidance corresponded closely with contemporary East India Company instructions to its own supercargoes, which typically epitomized the Chinese as "a very cunning subtle people."[28] Moreover, Ovington identified the tea plant's hardy physiognomy in withstanding cold winters, and the corresponding opportunity for its transplantation closer to home. "[M]ight it therefore be convenient to have it brought hither, there is nothing in the Nature either of our Ground or Air that seem to contradict its Increase among us," the writer wonders—although he is also alert to the practical and political obstacles confronting such an endeavor.[29] Ovington's combination of armchair bioprospecting and his eulogizing of tea's consumer benefits once again earned the commendation of the company directors, who assigned him a further gift of three pounds following the *Essay's* publication.[30]

Ovington's *Essay upon Tea* was the earliest substantial and original publication on the topic in English, but his knowledge of China was entirely secondhand. The first British naturalist to inspect tea growing in its native habitat, to collect botanical specimens for distribution in Europe, and to observe indigenous processes of manufacture was the Scottish surgeon James Cuninghame (d. 1709). Having already voyaged at least once to the East Indies in the mid-1690s, Cuninghame spent six months at Amoy (Xiamen) as part of an interloping venture between 1698 and 1699. In 1700, he accepted a permanent position with the New East India Company at its proposed English factory on Chusan (Zhoushan), an island port in the East China Sea servicing nearby Limpo (Ningbo).[31] Commercially the enterprise was disastrous, and Cuninghame himself seems to have endured financial disappointment, but during his stay of two and a half years he achieved unprecedented access to the local landscape and its botanical riches.[32]

Already, in the vicinity of Amoy, Cuninghame had noted in a field book his identification of "Tea, Planta floribus et foliis serratis Urticus subtus albicantibus" ("Tea, a flowering plant, with leaves serrated like nettles and whiteish underneath").[33] Now on Chusan, he "found the Tea grow plentifully on the tops of the Hills in small shrubs amongst the Pines," as he wrote home to the London apothecary James Petiver (c. 1665–1718). More importantly, he was able to take cuttings for his friend that he included within "a quare of Paper containing about 150 Specimens of different Plants." These he sent home on the *Macclesfield*, a New Company galley which departed Chusan in December 1700, under the private care of its surgeon Mr. Corbet.[34] When the "Specimens" reached Petiver in London on July 12, 1701, he judged the package "highly acceptable" and remarked particularly upon the "crenated Leaves" of the "Thea."[35] Later in 1701, Cuninghame dispatched to the prominent physician Hans Sloane an account of the cultivation of tea based on a year of observing the island's ecological and agricultural cycle:

> The 3 Sorts of Tea commonly carryd to England are all from the same plant, only the Season of the Year & the Soyl makes the difference. The Bohe (or Voüi, so calld of [the Wuyi] Mountains in the Province of Fokien, where it is chiefly made) is the very first bud gathred in the beginning of March, & dryd in the shade. The Bing Tea is the second growth in April: & Singlô the last in May & June, both dryd a little in Tatches or Pans over the fire. The Tea Shrub being an Evergreen is in Flower from October to Januarie, & the Seed is ripe in September & October following, so that one may gather both flowers and seed at the same time.[36]

Cuninghame was the first British traveler to observe, document, and collect Chinese plants in any meaningful detail. On Chusan he bemoaned how "the Jealousie of these People among whom we live restrains so much that we have no freedome of rambling," and regretted "the want of Chinese Physicians here which I had at Emuy [which] hinders me from giving you such informations as I could willinglie wish for."[37] Nevertheless, both covertly and diplomatically he was able to obtain sustained opportunities for natural historical enquiry. In relation to tea, Cuninghame resolved an enigma concerning the botanical source of darker "Bohe" tea and greener "Bing" and "Singlô." They "are all from the same plant," he determined, the differences between them arising from the season and method of preparation (Bohea oxidizes when it is "dryd in the shade," whereas the verdure of Bing and Singlô is preserved by its immediate firing in "Tatches or Pans"). Today botany similarly recognizes a single species of tea, *Camellia sinensis* (Figure 1.3). The scientific value of Cuninghame's fieldwork was brought to public attention when excerpts from his letters to Sloane (including

FIGURE 1.3. Specimen of tea (*Camellia sinensis*) collected by James Cuninghame on the island of Chusan (Zhoushan), circa 1701, Natural History Museum (London), Sloane Herbarium. © The Trustees of the Natural History Museum.

his observations on tea) were printed in the Royal Society's *Philosophical Transactions* in 1702, while the following year "Thea CHINENSIS vera potulenta" ("the true, drinkable CHINESE tea") was included in a catalog of specimens prepared on Cuninghame's behalf in the same publication by Petiver.[38] Around the same time,

Petiver commissioned an engraving of Cuninghame's tea cutting for the third "decade" of his *Gazophylacii Naturae et Artis* ("Treasury of Nature and Crafts").[39] Visually, textually, and materially, therefore, Cuninghame gratified the public desire for authentic knowledge about tea: a plant that remained decidedly exotic, even as its product became increasingly familiar.

The extraordinary efforts of James Cuninghame meant that improved natural historical and agricultural understanding of Chinese tea was made available to European botanists, both through Petiver's publications and—for those sufficiently well connected—in person at Petiver's domestic museum (purchased and absorbed within Sloane's much larger collection following the apothecary's death in 1718). Yet Cuninghame was lost at sea in 1709 before he was able to return home and publish a more complete Chinese herbal, while direct access to China's landscape and its resources remained heavily circumscribed for Europeans. The great Swedish botanist Carl Linnaeus (1707–1778), for example, labored for many years to acquire from China his own living tea shrubs to cultivate at his botanical garden in Uppsala. Once again, science and commerce collaborated as the professor networked and strategized with merchant travelers to East Asia in order to establish a reliable means for transporting seeds or saplings. When Captain Carl Gustav Ekeberg finally landed a number of tea plants on Swedish soil after a long voyage in 1763, Linnaeus rhapsodized, "Truly if it is Tea, I shall make your name, Mister Captain, more eternal than Alexander the Great . . . for God's sake, for your love of your Fatherland, for the natural sciences, and for all that is holy and famous in the world, treat them with the most tender care."[40] Although not all of Ekeberg's tea made it to Linnaeus alive, he was able to nurse at least one specimen to health. Yet Linnaeus could not bring it to flower, and by 1765 he despaired of ever being able to do so—a rare horticultural disappointment which meant that he was forced to rely on other authorities when formulating a botanical description of tea. As a consequence, a dissertation on "Potus Theae" ("The Tea-Drink"), defended by Linnaeus's student Pehr Cornellius Tillaeus on December 7, 1765, wrongly outlined two separate species, *Theæ Bohea* and *Theæ viridi* (Bohea and Green Tea) some fifty years after Cuninghame had settled the question to the contrary.[41] The formidable scientific gravitas of Linnaeus had the effect of promulgating this error among botanists during subsequent decades.[42] It was almost as if Chinese tea was actively resisting the disciplinary incursions of those European minds that attempted to call it to order.

Linnaeus's desire for tea was not merely a remote botanist's urge to classify the world. In tea he detected a cash crop that might enrich Swedish instead of Chinese landholders. To master the tea plant—botanically, horticulturally, proprietorially—would be to own the means of tea production, wresting it from farmers in East Asia.[43] This ambition Linnaeus nurtured and shared in correspondence with the British naturalist John Ellis (c. 1710–1776) over a period of around fifteen years. During the 1760s, Ellis became increasingly committed to a mode of global economic botany that sought to appropriate potentially profitable plants from regions beyond British jurisdiction, principally the Indian Ocean world, and to deploy them agriculturally in the colonial territories of North America and the Caribbean (he earned salaries as king's agent for West Florida from 1763 and as agent in London for Dominica from 1770). Linnaeus had actively encouraged these inclinations: as early as December 1758, he urged Ellis to see whether he could acquire "a living plant of the Tea" from China, for "I am very sure that this plant would bear the open air in England, as it thrives at Pekin, where the cold is more intense than in Sweden."[44] Acting on this prompt, Ellis wrote the following month to Thomas Fitzhugh (1728–1800), an East India Company factor bound for China, with instructions for securing and preserving tea seeds to send home to Britain.[45] By December 1760 a small shipment duly reached him in London from Limpo (the major coastal city near the island of Chusan). Ellis distributed the seeds not only to Uppsala, but also to "to each of our governors of provinces from New England to Georgia; so that I hope to establish Tea in America, by that means, in time."[46] Although none of this early batch was to germinate, Ellis's obsession was fired.

Throughout the 1760s Ellis remained in contact about tea with both Linnaeus and Fitzhugh, ruminating with the former about botanical principles and horticultural methods that would optimize the chances of receiving or rearing live plants in Britain, while directing the latter accordingly. After some false starts—and notwithstanding Linnaeus's gift from Ekeberg—on August 19, 1768, Ellis triumphantly wrote to Uppsala that "there is a tea tree brought over by Mr. Fitzhugh who has resided as the East India Company factor many years in China." In addition—and countering the thesis of Potus Theae—Fitzhugh confirmed to both Ellis and Daniel Solander (Linnaeus's protégé in London) what Cuninghame had already known. "[T]here is but one Species of the Tea Tree in china," Fitzhugh had deposed, for the variations between "Green & Bohea are owing to the difference of Soil, & time of gathering & drying it."[47]

Fitzhugh's first tree seems to have been "an old one" that had been potted up in China.[48] However, details in a publication that Ellis issued shortly thereafter, *Directions for Bringing over Seeds and Plants from the East-Indies . . . in a State of Vegetation* (1770), fleshed out a careful process developed with Fitzhugh and others that enabled more extensive transplantations. It demonstrates a fusion of Asian and European technologies that must have required careful liaison between British merchants and local agents in China. First, "tea-seeds" were gathered in "their pods or capsules" and taken "fresh from the tea-country at the latter-end of the year, to Canton, at the time that our East-India ships are preparing to depart." Second, in Canton the seeds were packed "into pound or half-pound canisters made of tin and tutenague . . . lined with silk paper" (tutenag was a malleable and largely nonferrous alloy, and like "silk paper" a traditional Chinese commodity). Third, en route to England, these seeds "shot out roots, owing to the heat of the climates they had passed through, and the confined moisture." While some were to be left in their "canisters" and planted upon arrival in England, producing a success rate of around 10 percent, others were sown on board the ship (Ellis's *Directions* animadverts extensively on the appropriate time, containers, and soil types for doing so).[49] Although Fitzhugh endured the agony of finding the maritime tea trees he raised in this manner "destroyed by rats" at St. Helena, others experienced greater success. In November 1769 Ellis consequently boasted to Linnaeus that "by this time twelve-month we shall have many hundred plants of the true tea growing in England."[50]

While Ellis undoubtedly possessed a botanical interest in the tea plant, his primary aim was to inaugurate large-scale Western plantations. In February 1772 he assured William Tryon, newly appointed governor of New York (and later a controversial Loyalist commander), that "[i]t is my ardent wish before I die to see the Tea tree of China established in North America. I am perswaded from the variety of Climates soils & Situation it may be propagated there."[51] To maximize both technical understanding of tea cultivation, and the number of seeds and plants available for export to the colonies, Ellis distributed his new arrivals to the most able professional gardeners in London. Fitzhugh's tree went to James Gordon's nursery in Mile End, while later instances were forwarded to James Lee at the Vineyard in Hammersmith and to William Aiton, director of the Royal Botanic Garden at Kew.[52] It must have been via one of these horticultural channels that a thriving shrub came to flourish at Syon House, the London estate of Hugh Percy (c. 1712–1786), Duke of Northumberland. When it bloomed in October 1771, the *London Evening Post* credited it as "the first that ever flowered in Europe," and optimistically envisaged that "we may soon have tea of our own production, and save some of our silver."[53] Northumberland's gardens became the resort of the philosophically curious: the pre-eminent botanical illustrator John Miller (1715–1792) drew and engraved his tree for publication, while the society physician John Coakley Lettsom (1744–1815) formulated a precise anatomical description of the plant to accompany his own pharmacological analysis of tea as medicinal beverage in *The Natural History of the Tea-Tree* (1772) (Figure 1.4).[54] Yet again, commerce and natural science cooperated to produce new

FIGURE 1.4. John Miller, "Green Tea, Publish'd according to Act of Parliament, Dec. 10. 1771." Hand-colored printed frontispiece to John Coakley Lettsom, *The Natural History of the Tea-Tree, with Observations on the Medical Qualities of Tea, and Effects of Tea-Drinking* (London: Edward and Charles Dilly, 1772), Wellcome Library (London). Courtesy of the Wellcome Library, London, under the terms of the Creative Commons Attribution (CC BY) license.

knowledge about tea—indeed, new tea itself—although by now the Chinese dimensions of this nexus were largely submerged in the global West beneath European modes of understanding and practice.[55]

Ellis's attempts at the intercontinental mediation of tea—from China to the American colonies via Great Britain—depended above all upon an embedded and sympathetic supplier at Canton, where the majority of the East India Company's East Asian trade was concentrated. After Fitzhugh had returned home, Ellis enlisted a succeeding resident factor and aspiring young botanist, John Bradby Blake (1745–1773).[56] Not only were Blake's abilities and enterprise unstintingly eulogized in Ellis's publications, but his own little-studied archive reveals an assiduous commitment to natural history, and above all to collecting and disseminating economically exploitable plant expertise.[57] In letters to his father (and Ellis's friend), Captain John Blake of Parliament Street, Blake outlines his ambition "to discover *All* Forrest Trees, Shrubs, Plants . . . useful in Medicine, Food, Dying or for Mechanical Uses, as well as Articles of Trade, that this Empire produces," for "I shall be happy, if my researches into the Natural History of this Empire, may prove of any benefit to my Native Country."[58] While these "researches" extended beyond tea—notably to include rice, indigo, and the tallow tree—this prime article of the East India trade fell well within Blake's province. His notes show that in 1771 he cultivated tea from seeds at his apartment in Canton, regularly documenting their development. Furthermore, Blake employed "a very ingenious Chinese" to undertake detailed botanical illustrations—following European conventions and "copying exactly from Nature by my assistance"—at various stages of the plant's growth.[59] Neither did he omit to communicate to his father what he learned, transmitting his advice that "this valuable Shrub" should be "planted early in April in a Green House not a hot house . . . in common Mould about One Inch (or rather less) deep. Three Seeds together in each hole."[60] As well as his own observations, Blake's manuscripts contain a knowledgeable quarto of twenty-four pages titled "Extracts concerning Tea," a paper that

FIGURE 1.5. Mauk-Sow-U (?), *E Chaw*, 1772, colored drawing, Oak Spring Garden Foundation, Upperville, Virginia. Image courtesy of the Oak Spring Garden Foundation, osgf.org.

was later printed in the East India Company's *Oriental Repertory*, where it was credited to his forbear at Canton (and by Blake's time, a company director) Frederick Pigou (1711–1792).[61]

While Ellis, Fitzhugh, and Blake had concocted means suitably ingenious to resource the growing of tea in Britain, their attempts to direct seeds onward to America met failure. Eighteenth-century Europeans were ultimately unable to procure from China the quality of horticultural supply or the detail of agricultural expertise required to transplant tea successfully across international contexts—a reminder that transculturation results in depreciation and loss as well as instantiating processes of transformative acquisition. James Clark (1737–1819), a Scottish physician in Dominica, George Young (d. 1803), superintendent of the Botanic Gardens in St. Vincent, Alexander Garden (1730–1791), a Charleston naturalist, and John Moultrie (1729–1798), deputy governor of East Florida, all confessed by letter during the second half of 1774 that the tea seeds they had received through Blake's industry had not vegetated (potential crops of rice and tallow trees looked more promising).[62] Any chance of further momentum was obliterated shortly thereafter by a combination of Blake's and Ellis's deaths and the outbreak of military hostilities in the American colonies (the causes of which were epitomized, ironically, by a dispute concerning the duties to be levied—or not—on tea shipped via Britain from China).[63]

Back in London, the East End nurseryman so trusted by Ellis repeatedly advertised the availability of "the true green Chinese Tea-Tree . . . now to be sold at James Gordon's Nursery-Garden at Mile-End, near Bow."[64] One hundred and fifteen years after Thomas Garway had tempted wealthy urbanites with an extraordinary dried leaf that had since become a dietary staple of ordinary laboring households, Gordon temporarily reassigned an exotic allure to tea. At the Mile End nursery, and thence across the properties of Gordon's moneyed clients, tea anticipated the wave of Chinese botanical introductions that would gradually transform European gardens. As a symbolic act within tea's ongoing transculturation, Gordon's commodification of the shrub at once epitomized the ongoing consolidation of Britain's powerful reach across the global East (just as it was receding in the West), while simultaneously signaling the absence of British control over a core component of its material basis (in the form of actual tea plantations in China). Complete economic naturalization would have to wait until the middle quarters of the nineteenth century: ironically, fascinatingly, and deeply problematically triangulating new series of oriental networks within the imperial tea plantations of India.

Notes

1. These themes are explored at length in Markman Ellis, Richard Coulton, and Matthew Mauger, *Empire of Tea: The Asian Leaf that Conquered the World* (London: Reaktion, 2015). Parts of this essay draw significantly on research completed for that book, and I am indebted to my coauthors for sharing their findings and expertise.

2. [Joseph Addison], *Spectator* 10 (March 12, 1711), in Joseph Addison and Richard Steele, *The Spectator*, ed. Donald F. Bond (Oxford: Clarendon, 1965), 1:44; Commissioners of Excise, "Observations on Memorial and Petition from Tea Merchants, which Suggested Possible Ways of Combating Widespread Smuggling" (1778), Treasury Papers, T 1/542, fol. 229r, National Archives, Kew.

3. On British living standards during the eighteenth century, see, for example, Charles Feinstein, "Pessimism Perpetuated: Real Wages and the Standard of Living in Britain during and after the Industrial Revolution," *Journal of Economic History* 58 (1998), 625–658; Maxine Berg, "Consumption in Eighteenth- and Early Nineteenth-Century Britain," in *The Cambridge Economic History of Modern Britain*, ed. Roderick Floud and Paul Johnson (Cambridge: Cambridge University Press, 2004), 1:357–386. On the history of a "consumer-oriented society," see Neil McKendrick, John Brewer, and J. H. Plumb, *The Birth of a Consumer Society: The Commercialization of Eighteenth-Century England* (London: Europa, 1982); the editors' introduction to *Consumption and the World of Goods*, ed. John Brewer and Roy Porter (London: Routledge, 1993), 1–15; Maxine Berg, *Luxury and Pleasure in Eighteenth-Century Britain* (Oxford: Oxford University Press, 2005).

4. Fernando Ortiz's concept of "transculturation" appeals precisely because it encodes processes of both cultural acquisition and loss: see his *Cuban Counterpoint: Tobacco and Sugar*, trans. Harriet de Onís (New York: Alfred A. Knopf, 1947), 97–103.

5. On the contemporary practice of "bioprospecting," see for example Londa Schiebinger, *Plants and Empire: Colonial Bioprospecting in the Atlantic World* (Cambridge, MA: Harvard University Press, 2004).

6. Lu Yu, *The Classic of Tea: Origins and Rituals*, trans. Francis Ross Carpenter (Hopewell, NJ: Ecco, 1974), 115.

7. Victor H. Mair and Erling Hoh, *The True History of Tea* (London: Thames and Hudson, 2009), 30; David Hawkes, ed. and trans., *The Songs of the South: An Ancient Chinese Anthology of Poems by Qu Yuan and other Poets* (Harmondsworth: Penguin, 1985), 269–270.

8. Gaspar da Cruz, *Tractado em quese co[m] tam muito por esteso as cousas da China* (Évora: Andre de Burgos, 1570), translated in Samuel Purchas, *Purchas his Pilgrimes* (London: William Stansby for Henrie Featherstone, 1625), 3:180.

9. "De l'Usage du Tay, qui est fort ordinaire en la Chine" ("On the use of tea, which is commonplace in China"), in Alexandre de Rhodes, *Divers Voyages et Missions du Père Alexandre de Rhodes en la Chine* (Paris: Sebastien and Gabriel Cramoisy, 1654), 62–67. My translations of de Rhodes's observations are available at https://qmhistoryoftea.wordpress .com/resources/de-rhodes-on-tea/.

10. Johan Nieuhof, *An Embassy from the East-India Company of the United Provinces, to the Grand Tartar Cham Emperour of China*, trans. John Ogilby (London: printed by John Macock for the Author, 1669), 127.

11. See Heren XVII to Anthony van Diemen, January 2, 1637, Amsterdam: quoted in G[eorge] Schlegel, "First Introduction of Tea into Holland," *T'oung Pao*, Second Series, 1 (1900), 468–472, at 469.

12. *Mercurius Politicus*, September 23, 1658.

13. Thomas Rugg, *The Diurnal of Thomas Rugg, 1659–1661*, ed. William L. Sachse, Camden Third Series 91 (London: Royal Historical Society, 1961), 10.

14. Thomas Garway, *An Exact Description of the Growth, Quality, and Vertues of the Leaf Tee, alias Tay* ([London], [c. 1664]). See also Markman Ellis et al., eds., *Tea and the Tea-Table in Eighteenth-Century England* (London: Pickering and Chatto, 2010), 2:1–3.

15. Ethel Bruce Sainsbury, ed., *A Calendar of the Court Minutes etc. of the East India Company, 1664–1667* (Oxford: Clarendon, 1925), 70.

16. "Saved to the Consumers of Tea by Taking off the Excise Duties" (c. 1783), Chatham Papers (1786–1800), PRO 30/8/294, fol. 168r, National Archives, Kew.

17. Thomas Short, *A Dissertation upon Tea* (London: W. Bowyer, 1730), 3; John Wesley, *A Letter to a Friend, Concerning Tea*, 2nd ed. (Bristol: Felix Farley, 1749), 4.

18. For East India Company tea imports before 1739, see [Robert Wissett], *A View of the Rise, Progress and Present State of the Tea Trade in Europe* (London: n.p., [1801?]); for the period after 1740, see the *Returns showing the Number of Pounds Weight of the different Varieties of Tea sold by the East India Company, for Home Consumption, in each Year from 1740 down to the Termination of the Company's Sales, together with the Average Prices at which such Teas were sold* (London: House of Commons Parliamentary Papers, 1845).

19. Nahum Tate, *Panacea: A Poem upon Tea: In Two Canto's* (London: J. Roberts, 1700), 16, 34.

20. [Richard Steele], *Spectator* 4 (March 5, 1711), in *The Spectator*, ed. Bond, 1:21.

21. [Daniel Defoe], *The Fortunes and Misfortunes of the Famous Moll Flanders &c* (London: W. Chetwood and T. Edling, 1721), 80.

22. The relationship between discourse and subjection is an overarching theme of Michel Foucault's writings; a succinct account can be found in his lecture "The Order of Discourse," in *Untying the Text: A Post-Structuralist Reader*, ed. Robert Young, trans. Ian McLeod (London: Routledge, 1981), 51–78.

23. See for example Richard H. Grove, *Green Imperialism: Colonial Expansion, Tropical Island Edens and the Origins of Environmentalism, 1600–1860* (Cambridge: Cambridge University Press, 1995); Richard Drayton, *Nature's Government: Science, Imperial Britain, and the "Improvement" of the World* (New Haven, CT: Yale University Press, 2000); Vinita Damodaran, Anna Winterbottom, and Alan Lester, eds., *The East India Company and the Natural World* (Houndmills: Palgrave Macmillan, 2015); Anna Winterbottom, *Hybrid Knowledge in the Early East India Company World* (Houndmills: Palgrave Macmillan, 2016).

24. The museum of Sir Hans Sloane is perhaps the most instructive instance: see James Delbourgo, *Collecting the World: Hans Sloane and the Origins of the British Museum* (London: Allen Lane, 2017).

25. John Ovington, *A Voyage to Suratt, in the Year, 1689: Giving a Large Account of That City, and Its Inhabitants, and of the English Factory there* (London: Jacob Tonson, 1696), 306–309.

26. East India Company Court Minutes, April 16, 1697: India Office Records, B/41, p. 337, British Library, London. The payment of £25 also covered the procurement and transportation from Ireland to Surat of "two Wolf Dogs."

27. John Ovington, *An Essay upon the Nature and Qualities of Tea* (London: R. Roberts, 1699), 14–15, 22–23.

28. "Com[m]ission & Instructions given by the Court of Directors of the English Company Tradeing to the East Indies To . . . The Councill for the Affaires of the said Company in China" (November 23, 1699): India Office Records, E/3/94, fol. 58v, British Library, London.

29. Ovington, *Essay upon . . . Tea*, 6.

30. East India Company Court Minutes, July 7, 1699: India Office Records, B/43, p. 35, British Library, London.

31. This was a period of institutional flux for the East India Company. In 1694 Parliament permitted noncompany merchants (or "interlopers") to undertake commercial voyages east of the Cape for the first time since 1600; while in 1698 a second ("New") East India Company was legally incorporated. The two bodies merged within a few years, and operated a monopoly from 1708 as the United Company.

32. For fuller accounts see Charles E. Jarvis and Philip H. Oswald, "The Collecting Activities of James Cuninghame FRS on the Voyage of Tuscan to China (Amoy) between 1697 and 1699," *Notes and Records of the Royal Society* 69 (2015), 135–153; and Jane Kilpatrick, *Gifts from the Gardens of China: The Introduction of Traditional Chinese Garden Plants to Britain, 1698–1862* (London: Frances Lincoln, 2007), 34–48.

33. James Cuninghame, "Plants of China etc" (1698–1699): Sloane 2376, fol. 63r (translation mine), British Library, London.

34. James Cuninghame to James Petiver, December 20, 1700, Chusan: Sloane 3321, fol. 65r–v, British Library, London.

35. James Petiver to James Cuninghame (draft), July 16, 1701, London: Sloane 3334, fols. 78r–79v (fol. 78r), British Library, London.

36. James Cuninghame to Hans Sloane, November 22, 1701: Sloane 4025, fols. 92–93, British Library, London.

37. James Cuninghame to James Petiver, November 22, 1701, Chusan: Sloane 3321, fol. 89r; and August 26, 1702, Chusan: Sloane 3321, fol. 100r, British Library, London.

38. James Cuninghame, "Part of Two Letters to the Publisher from Mr James Cunningham, F.R.S. and Physician to the English at Chusan in China," *Philosophical Transactions of the Royal Society of London* 23 (1702–1703), 1201–1209; James Petiver, "A Description of some Coralls, and other Curious Submarines . . . as also An Account of some Plants from Chusan and Island on the Coast of China; Collected by Mr James Cuninghame, Chyrurgeon & F.R.S.," *Philosophical Transactions of the Royal Society of London* 23 (1702–1703): 1419–1429.

39. James Petiver, *Gazophylacii Naturæ et Artis Decas Tertia* (London: Samuel Smith and Christopher Bateman, 1704) 33–34 and tab. XXI; James Petiver, *Catalogus Classicus et Topicus* (London: Christopher Bateman, 1709), 3.

40. Carl Linnaeus to Gustav Ekeberg, August 18, 1763, Uppsala: quoted and translated in Lisbet Koerner, *Linnaeus: Nature and Nation* (Cambridge, MA: Harvard University Press, 1999), 138.

41. Carl Linnaeus and Pehr Cornelius Tillaeus, *Potus Theae* (Uppsala, 1765), 2.

42. For further detail see Ellis, Coulton, and Mauger, *Empire of Tea*, 102–103, 108–110.

43. Koerner, *Linnaeus*, 135–139.

44. Carl Linnaeus to John Ellis, December 8, 1758, Uppsala, in James Edward Smith, ed. and trans., *A Selection of the Correspondence of Linnæus, and Other Naturalists* (London: Longman et al., 1821), 1:109–110. Letters from Linnaeus to Ellis were written in Latin.

45. John Ellis to Thomas Fitzhugh, January 29, 1759, London (draft): John Ellis Note-Book 2, fol. 3r, Linnean Society of London.

46. Ellis to Linnaeus, [December 1760], London, in Smith, *Correspondence of Linnæus*, 1:138–139.

47. Ellis to Linnaeus, August 19, 1768, London (draft): John Ellis Note-Book 2, fols. 37v–38r, Linnean Society of London.

48. Ellis to Linnaeus, August 19, 1768, London, in Smith, *Correspondence of Linnæus*, 1:229–235.

49. John Ellis, *Directions for Bringing over Seeds and Plants from the East-Indies and Other Distant Countries, in a State of Vegetation* (London: L. Davis, 1770), 3–8.

50. Ellis to Linnaeus, November 27, 1769, London, in Smith, *Correspondence of Linnæus*, 1:242–243.

51. Ellis to William Tryon, February 4, 1772, London (draft): John Ellis Note-Book 2, fols. 109v–110r, Linnean Society of London.

52. Ellis to Linnaeus, August 19, 1768, London, and September 25, 1770, London, in Smith, *Correspondence of Linnæus*, 1:229–235, 249–252.

53. *London Evening Post*, October 22–24, 1771.

54. John Coakley Lettsom, *The Natural History of the Tea-Tree, with Observations on the Medical Qualities of Tea, and Effects of Tea-Drinking* (London: Edward and Charles Dilly, 1772). Miller's engraving formed the frontispiece to the first edition, which was dedicated to the Duke of Northumberland.

55. On Lettsom's text, see my accounts in Ellis et al., *Tea and the Tea-Table*, 2:137–139; and Ellis, Coulton, and Mauger, *Empire of Tea*, 106–107.

56. For Bradby Blake's initial appointment, see India Office Records, E/4/863, p. 427, British Library, London.

57. See, for example, John Ellis, *Some Additional Observations on the Method of Preserving Seeds from Foreign Parts, for the Benefit of Our American Colonies* (London: W. Bowyer and J. Nichols, 1773), 14.

58. John Bradby Blake to John Blake, c. December 1772, Canton (extract): M–152 (Descriptions of Plants and Autograph Letters), not foliated, Oak Spring Garden Foundation, Upperville, VA. I am grateful to Peter Crane for enabling my access to digitized versions of the John Bradby Blake papers at Oak Spring.

59. John Bradby Blake, "E Chaw": M–152 (Index to the Second Volume of China Botanical Illustrations), not foliated, Oak Spring Garden Foundation, Upperville, VA.

60. John Bradby Blake, undated note: M–152 (List of Seeds and Plants Sent to England from China), not foliated, Oak Spring Garden Foundation, Upperville, VA.

61. "Extracts concerning Tea &ca &ca": M–152 (Seeds and Plants), not foliated, Oak Spring Garden Foundation, Upperville, VA; "Of Tea. Collected by Frederick Pigou Esq. at Canton, in 1753," in Alexander Dalrymple, ed., *Oriental Repertory* (London: East India Company, 1791), 2:285–300.

62. Letters to John Ellis respectively dated and addressed July 24, 1774, Dominica; July 30, 1774, St Vincent; September 13, 1774, Charleston; December 25, 1774, St Augustine: M–152 (Descriptions of Plants and Autograph Letters), not foliated, Oak Spring Garden Foundation, Upperville, VA.

63. On tea's role in the events remembered as "the Boston Tea Party," see Ben Dew's introduction to Ellis et al., *Tea and the Tea-Table*, 4:vii–xvi, and Ellis, Coulton, and Mauger, *Empire of Tea*, 203–208.

64. *Daily Advertiser*, November 2, 1772. Gordon placed a large number of similar advertisements for tea trees in metropolitan newspapers between 1772 and 1774.

2

China-Pugs

The Global Circulation of Chinoiseries, Porcelain, and Lapdogs, 1660–1800

STEPHANIE HOWARD-SMITH

In December 2011, a pair of porcelain pug dog figures manufactured in China during the reign of the Qianlong emperor (1736–1795) sold for $20,411, three times their estimated value, at an auction held by Christie's in Amsterdam.[1] Painted in *rouge-de-fer*, the dogs are modeled in a seated position with their heads twisted around at almost ninety degrees and their mouths agape. A single gilt bell hangs from each of their collars. On their bases, the pugs are marked with the sign of crossed swords—the identifying mark of the Meissen porcelain manufactory, based near Dresden in Saxony (Figure 2.1). These porcelain figures are tangible manifestations of a series of interlocking networks that shaped cross-cultural exchanges between East and West during the long eighteenth century. How did Chinese ceramicists come to imitate a porcelain figure in the fashion of a German manufacturer? To answer this question, my essay will consider the pug dog, both the living animal and its representations in porcelain and related ceramic media, as products of transculturation. I will trace the dogs' movements across continents and national borders and examine the broad

FIGURE 2.1. *A Rare Pair of Chinese Export Figures of Seated Pug Dogs,* Qianlong period (1736–1795), porcelain, 17.5 cm high, private collection. Photograph © Christie's Images / Bridgeman Images.

range of commercial trade networks, personal and social networks, as well as information networks involved in their distribution and in the transculturation of aesthetic ideas.[2]

During the long eighteenth century, lapdogs were strongly associated with chinoiserie artefacts.[3] The association of pug dogs with chinoiserie commodities was particularly pronounced. The pug became ubiquitous as a porcelain figure, eventually asserting itself as part of orientalized motifs both in the Chinese porcelain export market and in European ceramics. The prevalence of Chinese porcelain pugs may have contributed to the popular notion that the pug dog originated in China. Although there is no clear evidence to support this suggestion, this type of dog was coded as chinoiserie nonetheless, possibly because the living animal became so inextricably connected with its porcelain counterparts in the 1730s and 1740s.

Recent research has analyzed how both lapdogs and porcelain dog figures were associated with the Orient. Chi-ming Yang emphasizes the connections between the "toy dog and the porcelain dog, linked through association with the female consumer and distant Eastern origins," each demonstrating "the emergence of art valued as reproduction and the reproducibility of life valued as a form of art."[4] Yang notes that "the practices of petkeeping and china collecting developed in tandem in the female-oriented consumer culture of eighteenth-century England . . . as transportable yet delicate miniatures, dog and porcelain are particularly mobile emblems of exchangeability."[5] Yang's observation is underlined by a nineteenth-century satire that deliberately blurs the boundaries between porcelain pug and live animal. Describing a spinster offered at a marriage auction who "keeps three poodles, one china-pug, and a pair of the smallest spaniels ever beheld,"[6] and listing the china-pug between two other varieties of toy dogs, the author does not specify whether the china-pug is a live animal of Chinese origin or a porcelain figure. In other satires of the same period, lapdogs and their porcelain representations share similar objections as useless commodities. Writing in 1753 on simplicity in taste, Joseph Warton criticized "the extravagant lovers and purchasers of CHINA" who "spend their lives collecting pieces where neither perspective, nor proportion, nor conformity to nature are observed . . . no genuine beauty is to be found in whimsical or grotesque figures, the monstrous offspring of wild imagination, undirected by nature and truth."[7] Eighteenth-century arguments against chinoiseries, presenting them "at worst, [as] a deviant obsession, or, at best, a rather disreputable affectation,"[8] were also extended to live pugs.[9]

Just as eighteenth-century critics of Chinese porcelain emphasized the unnatural form of the chinaware, critics of lapdogs and their mistresses frequently questioned the unnatural character of the live animals and of lapdog ownership. Targets of criticism included the lapdog's mode of living with all the accoutrements of human luxury, the supposed intimacy between female owners and pets, and, particularly with regard to pugs, their unnatural appearance.[10] As early as 1753, artist Paul Sandby presented the pug as evidence of William Hogarth's aesthetic preference for "Deformity" following the publication of *The Analysis of Beauty* in the same year.[11] In 1829, zoologist James Wilson lamented the creation of the "puny and contemptible pug-dog," one of the "artificial creatures, incapable of subsisting without the aid of man." Wilson considered the pug dog not only to be unnatural, but also an affront to natural law itself: "it is clear from what we know of the harmonious laws which regulate the animal economy, that no such creature as a pug-dog could ever have existed in a state of nature."[12]

The Pug Dog

By the early nineteenth century, a recognizable variety of dog generally identified as a "pug dog" or "Dutch mastiff" had been bred in Europe for over a hundred years. Many of the theories put forth in nonacademic popular histories of the breed written by and for pug owners suggest that the dogs were imported from China by merchants of the Dutch East India Company, or by Portuguese merchants.[13] More recent research by canine geneticists suggests, on the contrary, that the pug is a European breed, originating in northwestern Europe. Nonetheless, the long association of pugs with China has left its own historical record.

European depictions of live pugs bear at least a superficial resemblance to some dogs featured in oriental art, for instance, the beige lapdog in a 1720 woodblock print by Japanese artist Okumura Masanobu.[14] Despite the assertions of most twenty-first-century pug fanciers, the dog's oriental heritage is far from certain. There was undoubtedly a long history of women keeping lapdogs in general in imperial China. Court ladies of the Tang dynasty (618– 907) playing with their lapdogs were depicted by artists such as Zhou Fang and modeled in earthenware carrying or accompanied by their pets, although it is not certain that any of these dogs were precursors to the pug.[15] Contradictory genetic analyses of twenty-first-century pugs suggest that, if the breed originated from China at all, there has been significant interbreeding with dog types not native to East Asia. In this respect pugs differ from several other dog breeds commonly identified as Chinese, including Chow Chows, which carry evidence of their ancient Asian origins, a so-called basal signature, in their genes. In the case of the pug "a significant degree of introgression with European breeds is recorded or strongly suspected."[16] A 2010 study of canine DNA also suggests that the pug has "discordant phenotypic/functional and genetic group assignments." Citing "evidence for breed admixture of Mastiff-like and toy breeds," vonHoldt and her colleagues even state that the examined DNA does not reflect an Asian background at all, but rather a European heritage.[17] However, a study from 2017 notes that "the pug dog groups closely with the European toy breed, Brussels griffon, in the toy spitz clade but also shares extensive haplotypes with the Asian toy breeds as well as many small dog breeds from multiple other clades. This likely indicates the pug's early exportation from Asia and subsequent contribution to many small breeds," suggesting that the pug may have originated in China.[18] Whatever conclusion specialists may reach, twenty-first-century pugs, the result of intensive selective breeding in the later nineteenth and twentieth centuries, differ to a considerable degree from their early modern counterparts.

It is unclear to what extent eighteenth-century fanciers of pugs considered the origin of the dogs to be Chinese. Neither do we know whether such knowledge might have helped to establish a cultural connection between pugs and porcelain, and whether this may have made the pug such an attractive subject for porcelain modelers. Many natural historians, including the Comte de Buffon, believed that the pug's origins lay in Europe.[19] Known in its earliest days in Britain as a "Dutch mastiff" or "Dutch dog," the pug retained its Dutchness throughout the long eighteenth century.[20] Hester Lynch Piozzi's much-quoted description of the "little pug dog or Dutch mastiff" as a "transplanted Hollander, carried thither originally from China," remained an exception among early modern sources in its explicit reference to the pug's oriental heritage.[21] An allusion to the pug's Chinese origins can also be found in Eliza Haywood's last major work, *The Invisible Spy*, which details the funeral arrangements for a Dutch mastiff, an animal the narrator nicknames "monsieur Le Chin."[22] Clearly there was a popular understanding that the pug was in some sense

Chinese. However, the natural history writer Richard Brookes identified the pug's origins not in China, but rather in Indonesia, writing that "the DUTCH MASTIFF or PUG DOG, seems to have some relation to a Bull Dog; but he is much less, and some of them are exceeding small . . . it is hard to say what they are designed for, for they pursue no game in these parts, which perhaps may be owing to their being brought from *Batavia* as some affirm; for there may be animals in that country which are never seen here."[23] This demonstrates that the pug was associated with Dutch trade in the Orient, if not with China specifically. It was only from the Victorian period onward that British zoologists and pet owners began to assert the pug's Chinese heritage more decisively. In "Culture in Miniature," Yang describes how nineteenth- and twentieth-century writers and dog owners cultivated a narrative around the pug dog's supposed Asianness and points out that such ascriptions were often racialized. Referencing zoologist John Edward Gray's 1867 comparison between the physiognomy of East Asians and that of the pug dog, Yang calls attempts to prove the dog's Asian origins "perhaps the most troubling legacy of mapped resemblances between the canine and the human."[24]

The Dog Trade

Dogs have been present in human societies since their early development, "partners in the crime of human evolution" as Donna Haraway terms their role in hominid/canid coevolution in *The Companion Species Manifesto* (2003).[25] Haraway notes that "these large-bodied, globally distributed, ecologically opportunistic, gregariously social, mammalian co-travelers have written into their genomes a record of couplings and infectious exchanges to set the teeth of even the most committed free trader on edge."[26] More specifically, dogs have been traded between human societies and cultures for millennia. The Greek geographer Strabo listed hunting dogs among pre-Roman Britain's primary exports.[27] By the eighteenth century, live dogs were exchanged internationally along the same land and sea routes between Europe and the Orient as ceramics and other material goods. A 1665 portrait of the family of Pieter Cnoll, Senior Merchant of the Dutch colony of Batavia, depicts the family alongside their European lapdog.[28] In his *Natural History of Selborne* (1789), Gilbert White mentions a "dog and bitch of the *Chinese* breed from Canton," probably Chow Chows, brought back to Britain by his "near neighbor, a young gentleman in the service of the *East-India* Company."[29] Dogs could also be presented as gifts between traders rather than as cargo. Dutch magnate David Leeuw (1682–1755) sent a dog along with bottles of wine from Amsterdam to cloth merchants in the English West Country.[30] Animals were also routinely exchanged between European envoys and the Chinese court and other officials. In a 1703 report to the East India Company Court of Directors, Allen Catchpole, the president of the China Council, described how "we are extremely plagued for Curiosities of Birds and Dogs" as presents for the Chinese port officials, noting that one agent of the company, John Dolben, had "paid the Measurage of the Ship with one great Irish Dog."[31] John Bell, a Scottish traveler accompanying a Russian delegation to Peking in 1720, observed the transfer of several animals between an ambassador and a Mandarin "as a present from himself to the Emperor," including "several toys of value, a fine managed horse, some grey-hounds, and large buckhounds."[32] Bell also observed that "the Chinese, in general, are very fond of little harlequin dogs that play monkey tricks." A type of black and white spotted lapdog was sometimes known as the "harlequin dog," and such dogs are depicted in eighteenth-century Chinese export paintings.[33] Because the black face of the pug reminded contemporaries of the mask of the commedia dell'arte character, one name for the dog in France was *carlin*, the nickname of the renowned harlequin actor Carlo Bertinazzo.

By the time Pieter van Hoorn led the third Dutch embassy to China (1666–1668), Chinese officials expected transactions to involve lapdogs. According to Olfert Dapper, seventeenth-century commissioners from Peking summoned van Hoorn's interpreter to ask, "if the *Hollanders* had no Blood-Corral, little Dogs, and other Trifles to sell."[34] The circulation of dogs between China and

Europe was clearly a mutual interest and occurred in both directions. Like other European writers on China, Dapper also observed the popularity of lapdogs more generally with Chinese women, usually as part of an exoticized description of their appearance and activities. Twice he comments on lapdogs in relation to the isolation in which Chinese wives were kept. He considers lapdog ownership a natural consequence of such a lifestyle: "the Women in their Privacies, to pass away the time, entertain themselves with little Dogs, Birds, and the like pleasers of Fancy."[35] Later Dapper observes that women "are seldom permitted to go out, except on extraordinary occasions, and are carry'd in Sedans made for that purpose, and so closely shut, that there is not the least Crevise through which they may be seen. Moreover, they spend their time in breeding little Dogs, Birds, and the like."[36] Dapper seems to regard the ownership of lapdogs as further evidence of a life defined by idleness and leisure. Indeed, this image of Chinese women was also disseminated through portraits of Chinese ladies created for the European export market that depicted the women alongside their pets, including lapdogs.[37] Writing from Canton in 1751, the Swedish East India Company chaplain Olof Toreen noted the high esteem of Chinese women for their toy dogs: "Their dogs can do no more than bark, little dogs especially. *Spanish* ones are the delight of the *Chinese* ladies; and their husbands pay dearly for them: and I think there is some husbandcraft in it; for the affections must be fixed on some object."[38] There is an almost erotic quality to Dapper's (and, to a lesser extent Toreen's) descriptions of the role of the lapdog in the lives of Chinese ladies, suggesting that their seclusion stimulated the imagination of male English observers. References to lapdogs follow descriptions of transportation in sedan chairs and foot binding; both women and toy dogs seem to be ornamental in purpose and design. The Jesuit scholar Athanasius Kircher also echoed Olfert Dapper when he noted the fondness of Chinese women for their lapdogs (*catulis*) in his compilation of the reports of Jesuit missionaries, known as *China Illustrata* (1667).[39] John Ogilby included translated extracts from Kircher's work in his 1673 translation of Johan Nieuhof's *An*

Embassy from the East-India Company of the United Provinces to the Grand Tartar Cham, Emperor of China, including the observations on Chinese women and dogs: "to pass away their Time, they sport with little Dogs, Birds, and such Delights."[40] Dapper's characterization of pet ownership as a means to fritter away idle hours of leisure agreed with Kircher's representation of Chinese customs in *China Illustrata*. Knowledge about lapdogs, including their geographical origins, appearance, value, owners, and their treatment within China, was exchanged in complex information networks. Reports by visitors to China were translated into English after their return to Europe; individual narratives were collated before being disseminated across the continent and translated into different languages. These translations were occasionally translated further into other languages via indirect translations, as in the case of Toreen: his "A Voyage to Suratte," for example, was included in Peter (Pehr) Osbeck's *Dagbok öfwer en ostindisk Resa åren*, published in 1757, which was then translated from Swedish into German by Johann Gottlieb Georgi in 1765, and then translated into English from Georgi's German edition by John Reinhold Forster in 1771.[41]

Visual depictions of China and its inhabitants, including their pet dogs, also circulated through these books and traveled across national borders. Both *China Illustrata* and Ogilby's translations of Nieuhof's and Kircher's texts are illustrated with an etching of the "Supreme Monarch of the China-Tartarian Empire," depicting the Kangxi emperor alongside a small dog (Figure 2.2). The animal, which has a reduced muzzle and a short coat, resembles similar pug-like dogs featured in seventeenth-century Dutch art, for instance those in Paulus Potter's *A Hall Interior with a Group of Eight Dogs* (1649). It also bears a likeness to the female dog cast in bronze by Dutch sculptor Hubert Gerhard around 1600; a lapdog pictured in August Erich's portrait of Maurice of Hesse-Kassel's family (1618–1628); or even the dog in the bottom right-hand corner of Hans Rottenhammer's 1603 painting *Minerva with the Muses on Mount Helicon*.[42] Hendrick Goltzius depicts similar dogs in his 1616 painting *Lot and His Daughters* and in his 1587 engraving of Justus Lipsius accompanied by his animal companion "Mopsus."[43]

The Supreame
MONARCH
of the CHINA-TARTARIAN Empire.

FIGURE 2.2. Wenceslaus Hollar after Johannes Nieuhof, "The Supreame Monarch of the China-Tartarian Empire," in *An Embassy from the East-India Company of the United Provinces, to the Grand Tartar Cham, Emperor of China, deliver'd by their excellencies Peter de Goyer and Jacob de Keyzer*, 1673, etching, Call # 141–515f, plate 1 after page 68, Folger Shakespeare Library, Washington, DC. Used by permission of the Folger Shakespeare Library.

While the print of the Kangxi emperor offers no evidence for the pug's Chinese origin, it demonstrates that pug-like dogs had been associated with the Orient since the mid-seventeenth century. Two works by the Italian artist Giovanna Garzoni (1600–1670) depicting small pug-like dogs next to Chinese porcelain bowls provide further evidence of an early connection between the dogs and China and, perhaps more importantly, between these dogs and porcelain.[44]

While many lapdogs traveled as commercial cargo or as official diplomatic gifts, others were exchanged as part of private, personal networks. For instance, Madame de Pompadour sent her pet pug to the Comte de Buffon's chateau at Montbard before she died.[45] Francis Coventry's popular 1751 it-narrative, *The History of Pompey the Little*, also reflects a widespread understanding of toy dogs' international movements and the degree to which these illuminate other human transactions: In *Pompey the Little*, the eponymous Bolognese lapdog is brought to England by a returning Grand Tourist as a memento of his relationship with Pompey's original owner, an Italian courtesan.[46] In *Animal Companions* (2015), Ingrid Tague notes that requests for and offers of pets in Horace Walpole's correspondence provide evidence of a "vast network of circulating animals, sometimes exchanged as signs of friendship, sometimes for more mercenary reasons. Like other gifts, they were frequently used to grease the wheels of patronage."[47] Walpole himself was pressed by his dying friend, Madame Deffand, to adopt not only her lapdog, Tonton, but also her collection of china.[48] Both living lapdogs and porcelain representations were exchanged in such transactions.

The Lapdog and Chinoiserie

Many literary texts of the period closely associated live lapdogs with other Asian luxuries, if only because both were coveted by the same group of women. In Jonathan Swift's *The Journal of a Modern Lady* (1728), the loud gossiping of ladies taking tea "set the very lapdog barking."[49] Connections between the pug specifically and oriental commodities are even more apparent in an anonymous poem published in the *Diverting Post* in 1705, describing how a married couple argues over the wife's pet pug.[50] While the wife praises her "voluptuous Pug, / My dub-nos'd, crop-ear'd, dainty Rogue," the husband opposes it. Playing on the English pseudo-Dutch phrase "hogan-mogan" to describe it, perhaps semi-ironically, as a "Hogan whelp," he threatens to "see you at the Devil, / E'er a Dutch Dog shall be my rival." Although the poet does refer to the Dutchness of the pug, the animal is also associated with other fashionable imports from East Asia. The pug owner compares its fur to oriental lacquerware and the dog itself plays with the lady's hand-fan, another fashion item imported from China:

> How smooth his Coat is, how Japan?
> See how he gnaws and paws my Fan;
> 'Tis pitty but thou was't a Man.

External similarities between the pug and the luxury oriental commodity, as well as its interaction with a Chinese accessory, are evidently pleasing to the wife. With her rhetorical question and imperative verb form she alerts her husband (as the anonymous poet alerts his readers) to the oriental aspects of the dog's appearance and behavior.[51] As in other satirical poems involving women and lapdogs, the pug is presented as a rival to the lady's husband; in this particular case, the dog ingratiates itself with its mistress in part through its connections with fashionable pieces of chinoiserie. Live pugs were also frequently compared to and associated with porcelain. In "The Insolence of Office" (1776), Richard Graves creates an extended similarity between a pug dog and porcelain vessels.[52] Graves compares the "dog call'd Pug . . . with nose erect and saucy air," the "curst tyrant of the parlour," to "a little man, / In stature not above a span; / In shape, much like a china jug." All three items—the pug, the man, and the jug—are united by their roundness; the man also has a "smooth and smug face," presumably much like the pug.

Graves was not the first eighteenth-century writer to associate a lapdog with porcelain; however, most other comparisons emphasize the similarities in treatment of the two rather than a physical resemblance. In the fourth edition of Alexander Pope's *The Rape of the Lock* (1715), for instance, the causes of Belinda's theoretical distress become increasingly inappropriate:

> Not louder Shrieks by Dames to Heav'n are cast,
> When Husbands, or when Lapdogs breath their last;
> Or when rich *China* Vessels, fall'n from high,
> In glittering Dust, and painted Fragments lie![53]

Literary representations of consumerist behavior as well as critical studies of women's consumerism during the long eighteenth century have long remarked upon connections between lapdogs and porcelain as luxurious, fashionable commodities.[54] Porcelain and lapdogs were also frequently depicted together in visual art, beginning in the seventeenth century with Giovanna Garzoni. More recent scholarship on lapdogs and pet ownership in the eighteenth century also explores the connections between lapdogs and porcelain. Both Yang and Tague discuss a quotation from Charles Molloy's *The Coquet* (1718) in which "Madamoiselle Fantast" declares that "the next thing I should take a Fancy to, may be either a Lap-Dog, a Husband, or a piece of *China*."[55] Tague observes that through the listing of china alongside lapdogs (and husbands), Pope and other eighteenth-century writers attacked "the fickleness of female affection, which might seize upon any object in worrisome failure to distinguish kind from kind."[56] An article published in a literary journal in 1821 mocked a spinster whose "parroquets and pug-dogs afford full exercise for her energies, until supplanted by old china and tortoiseshell cats."[57] Not unlike Graves, who presented the pug and the china jug as *physically* indistinguishable, later writers suggested that porcelain and lapdogs along with other luxury consumables were interchangeable in the eyes of their female owners. Live lapdogs could also become users of porcelain themselves. In Robert Dodsley's *The Toy-Shop* (1744, 11th ed.), the master of the eponymous establishment presents potential customers with stuffed little dogs that "never eat but upon Plate and

China."[58] In Haywood's *The Invisible Spy*, monsieur Le Chin's living peers drink from "four china soup bowls, full of clear water," placed in each corner of their living quarters.[59] In Susan Ferrier's *Marriage*, published in 1818 but set in the late eighteenth century, Lady Juliana becomes ecstatic when she perceives the resemblance between some porcelain pug figures and her own pets: "Oh, the dear, dear little puggies! I must have them to amuse my own darlings. I protest here is one the image of Psyche; positively I must kiss it!"[60] Ridiculed as consumers of explicitly Asian luxury products and as fashionable but deformed and overvalued animal-commodities, oriental lapdogs serve to reinforce eighteenth-century critiques of female taste and consumerism in the West.

Meissen Pugs

Few porcelain animal figures were modeled in China before the growth of an export market during the seventeenth and eighteenth centuries, the period when Chinese porcelain manufacturers began to produce them specifically for European consumers. Some ceramic representations of dogs were made of earthenware during the Han dynasty (206 B.C.E.–220 C.E.), a period that pre-dates the European porcelain trade.[61] Court ladies with lapdogs were also modeled in earthenware by ceramicists of the Tang dynasty (618–907), although it is impossible to tell whether any of these sculptures reached European shores in the eighteenth century. Like the china figure as an art form in general, the trend for porcelain pugs can be traced back to the Meissen factory near Dresden in Saxony, the first and most prestigious of the European porcelain factories. First designed by model master Johann Joachim Kändler in May 1734, the Meissen pug became something of an unofficial symbol of the factory over the course of the eighteenth century.[62] By the beginning of the nineteenth century, approximately half of Meissen's porcelain dogs depicted pugs.[63]

The initial impetus for the production of pug dog figures at Meissen and Kändler's apparent proclivity for modeling pugs are as uncertain as is the origin of the live pug itself. It is possible that members of the *Mopsorden*—literally, the "Order of the Pug," a para-masonic organization founded in response to Pope Clement XII's bull of April 1738 banning Catholics from becoming freemasons—might have fueled a demand for pug figures to display their affiliation.[64] Kändler may also have hoped to appeal to the personal taste of one of the members of the *Mopsorden*, Heinrich, Count von Brühl, director of the Meissen porcelain manufactory, who was not only the most important politician in Saxony, but also a pug owner. Kändler played to von Brühl's fondness for these dogs in his choice of subjects, creating likenesses of the director's pugs and taking commissions for snuffboxes shaped as or featuring the dogs.[65] The stocky body, short and smooth coat, small size, and shortened muzzle of the pug as well as his curled tail, which could be affixed to the main body of the figure in two places, reducing the likelihood of breakage, may also have made the dog an appealing subject (Figure 2.3). By the 1740s and 1750s,

FIGURE 2.3. Johann Joachim Kändler (Meissen porcelain manufactory), *Pug Bitch with Puppy* and *Pug Dog*, circa 1741–1745, porcelain with enamel paint, 17.9×18.6×11.8 cm and 17.3×19.2×11.1 cm, PE 3894 and 3895, Porzellansammlung, Staatliche Kunstsammlungen Dresden. © Porzellansammlung, Staatliche Kunstsammlungen Dresden, Photo: Jürgen Karpinski.

the robust pug had become an established icon of the Meissen porcelain factory—almost a secondary logo in addition to the crossed swords. Meissen pugs proved to be as enticing to wealthy English consumers as they had already become to continental European customers. Brownlow Cecil, ninth Earl of Exeter, kept an inventory of his pugs in a day-book at Burghley House, Lincolnshire. Written between 1763 and 1777, his day-book lists "two pug dogs" among the "Dresden china on ye chimney" in the Japan Closet.[66] William Beckford, author of the orientalist-gothic novel *Vathek*, also owned "a pug dog, beautifully modelled in Dresden porcelane, and colored from nature."[67]

The Generic Porcelain Pug

From the mid-eighteenth century onward, new European manufactories began to model their own pug figures, capitalizing on the trend that Meissen had established and leaning on the Meissen pug's reputation as a superior product and on its status as a luxury consumer good. By the close of the eighteenth century, porcelain pugs were manufactured not only across the German states and kingdoms (at Ludwigsburg, Nymphenburg, Frankenthal, Crailsheim, Schrezheim, and Höchst), but also in France (Saint-Cloud, Vincennes, and Mennecy), Austria (Du Paquier and Vienna), Italy (Capodimonte) and England (Chelsea, Bow, Longton Hall, Derby, William Duesbury & Co., and Charles Gouyn's factory). Many of these models were imitations of original Meissen pieces to varying degrees of success. Among the most popular ones were pairs of male and female pugs, based on figures first modeled at Meissen between 1741 and 1745. Almost identical models of the latter were manufactured in France at Saint-Cloud and at Vincennes, and in England by the Lowestoft porcelain factory, the Longton Hall factory, and William Duesbury and Co.[68]

The influence of the Meissen porcelain pug also extended to the production of figures made from other, less prestigious materials such as earthenware, enamel, and stoneware. It affected, for instance, the design of manganese tureens in the shape of pugs from the German city of Erfurt, a pug-shaped faience bonbonnière from Pruszków in Poland (c. 1770), a faience figure from Belgium (late eighteenth century), a stoneware figure from Staffordshire (c. 1745–1755), and an enameled bonbonnière from Bilston in England (1760–1780), possibly modeled after a Meissen snuffbox.[69] The trademark posture of Meissen pugs—squatting on haunches, head turned awkwardly to the viewer—was not limited to porcelain imitations, but also affected the way live dogs were represented in various media, ranging from François Boucher's portrait *A Young Lady Holding a Pug Dog* (c. 1745) to the stuffed body of the duc d'Enghien's pug Mohiloff (c. 1804).[70]

From the mid-eighteenth century onward, Meissen pugs circulated in European commercial and private collectors' networks, both as desirable commodities in their own right and as models for imitation by other European porcelain manufactories. Once a Meissen pug (or one of its imitations) had found its way to China, it became an important factor in the production of "export porcelain," manufactured explicitly for European consumers. Traveling the same trade routes and benefiting from the same commercial networks that had brought Western pug models to Asia, orientalized porcelain lapdogs thereafter began to head in the opposite direction, returning from East to West.

Orientalized Lapdogs

Initially, porcelain dogs made in China may have imitated Delftware models.[71] But with pug dog figures being among Meissen's signature products, it was inevitable that, sooner or later, samples of the more prestigious models would also be brought to China for replication and profit in the export market. Porcelain models of pugs created in eighteenth-century China, including the pair from the Qianlong era described at the beginning of this essay, drew explicitly on Kändler's work. They exhibit the same slightly awkward sitting position as several genuine Meissen pug models and are painted in a European naturalistic manner.[72] Crucially, the undersides of the copycat figures even display an imitation of the crossed swords—Meissen's famous identifying marks. Several other models of porcelain pugs for the European market were also produced in China in a range of sizes, up to 42 cm (16.5 inches) tall.[73] These figures featured design details typical of Meissen pugs, such as bell collars, albeit more orientalized.[74] Chi-ming Yang has suggested that couples of male and female pugs for the export market evoked the Buddhist lion statues found outside Chinese Temples,[75] but it is more likely that the Chinese figures were copying Meissen's then iconic pairs of pug figures. Kändler himself may have had Buddhist lion statues in mind when he modeled the first such pairs in 1741. Moving between continents, the pug underwent a series of transformations: possibly from East Asia to Europe as a live dog; from Europe back to China as a Europeanized porcelain figure; and finally, back to the West as a more or less orientalized piece of Chinese export porcelain.

Just as the faux-Meissen Qianlong pugs, other eighteenth-century Chinese export pieces featuring different varieties of lapdogs also tended to copy specific European designs. The movements of the "Parrot and Spaniel" pattern, for example, illustrate the circulatory nature of cross-cultural influences on porcelain aesthetics. The pattern was based on the designs of Petrus Schenk, a Dutch engraver, and depicted a toy spaniel taunting a large parrot resembling a scarlet macaw chained to a wooden perch. First produced at Meissen in the 1740s, the parrot and spaniel pattern was later copied by Chinese porcelain manufacturers in Jingdezhen and exported back to the European market.[76] Surviving items from a Chinese export-ware parrot and spaniel pattern tea service in the Victoria and Albert Museum, London, made in the 1740s or 1750s, show the effect of this network of reinterpretations.[77] As with other examples of export porcelain, the design of the tea service was altered to fit the expectations of wealthy European consumers. To be instantly identifiable as Chinese in origin, it was orientalized. The vibrant enamel coloring of the Meissen plate was dropped in favor of blue and white ware, so desirable to European collectors of oriental porcelain. Almost recalling a dragon motif, the scrollwork resembles the curvature practiced by Chinese artisans more than the rococo style of the original pattern. The form of the parrot and the spaniel remain largely unchanged, although the execution is more Chinese in style. That the design was altered to fit a conspicuously oriental aesthetic suggests that the two animal representations were already in keeping with the chinoiserie scheme. The potentially Far Eastern origins of the spaniel and the exotic nature of the macaw (albeit indigenous to South America rather than Asia) reinforce the orientalist attraction of the service.

The lapdog remained a favored element in chinoiserie design during the eighteenth century. Even before Kändler first modeled a porcelain pug, Meissen porcelain painter Johann Gregorius Höroldt frequently incorporated lapdogs into exoticized chinoiserie scenes of mandarins going about their daily business.[78] Chinese export figures representing Dutch family groups, another popular genre for export to Europe, also featured lapdogs.[79] An early eighteenth-century group made of typical Dehua blanc de chine porcelain represents a Dutch couple surrounded by their children, a monkey, and a lapdog; a later-produced enamel-colored Jingdezhen figure incorporates only the lapdog.

By orientalizing European designs for their reintroduction to Europe, the Chinese export porcelain trade reinforced the close association between lapdogs, especially pugs, and chinoiseries. Porcelain manufactories in Germany and Britain followed suit and integrated lapdogs into explicitly oriental designs and motifs. A mug-and-cover set created in the Bow porcelain factory during the late 1750s incorporates a sitting pug as the handle of the cover (Figure 2.4).[80] The mug and its cover are decorated in a manner that imitates Japanese porcelain; the stylized floral motif and the color palette are typical of Kakiemon ceramicware. The drinking vessel is designed in such an explicitly oriental fashion that one might assume that the addition of the pug was regarded as compatible with the general style of the mug.

FIGURE 2.4. Cover, Bow porcelain factory, part of a set listed as *Mug and Cover*, circa 1755–1760, soft-paste porcelain, 20.9 cm high, 12.1 cm diameter, 414.109&/A–1885, Victoria and Albert Museum, London. Photograph © Victoria and Albert Museum, London.

Seventeenth-century travel accounts by Europeans about China often featured descriptions of small dogs as companion animals and hospitality gifts. Through this close association, Westerners came to perceive lapdogs, especially pugs, as part of Chinese culture, although more recent genetic evidence may suggest otherwise. Circulating the globe from East to West and from West to East, live pugs and their porcelain representations manufactured both in eighteenth-century Europe and in China contributed to a complex process of cultural exchange and mutual imitation. Overlapping commercial, personal, and aesthetic networks facilitated the transcultural success story of the pug. By the third decade of the nineteenth century, however, the popularity of the snub-nosed intercultural mediator had waned. British consumers began to look for porcelain dogs more unambiguously tied to the history of their own country and replaced "oriental" pugs with the spaniels of King Charles II.

Notes

1. *A Rare Pair of Chinese Export Figures of Seated Pug Dogs*, Qianlong period (1736–1795), porcelain, 17.5 cm high, private collection, Christie's Amsterdam (sale 2875, December 13–14, 2011; lot 529), http://www.christies.com/lotfinder/Lot/a-rare-pair-of-chinese-export-figures-5524008-details.aspx.

2. I thank my supervisor, Markman Ellis, for his advice and generous feedback on earlier drafts of this chapter.

3. For Samuel Johnson's definition of a "lapdog" as "a little dog, fondled by ladies in the lap," see *A Dictionary of the English Language* (London: J. & P. Knapton, 1755), 2:1168.

4. Chi-ming Yang, "Culture in Miniature: Toy Dogs and Object Life," *Eighteenth-Century Fiction* 25, no. 1 (2012), 139–174.

5. Yang, "Culture in Miniature," 153.

6. Louisa Henrietta Sheridan, ed., "An Auction for Wives," in *The Comic Offering; Or, Ladies' Melange of Literary Mirth for 1834* (London: Smith, Elder and Co., 1834), 8.

7. Joseph Warton, "No. 26," in *The World, for the Year One Thousand Seven Hundred and Fifty Three*, ed. Edward Moore [pseud. Adam Fitz-Adam] (London: R. & J. Dodsley, 1753), 155–156.

8. David Porter, *The Chinese Taste in Eighteenth-Century England* (Cambridge: Cambridge University Press, 2010), 20. "[T]he chinoiserie style seems consistently to have troubled this boundary between cultivated and vulgar taste, fine art and the fripperies of fashion. On the one hand, period satirists regularly denounce Chinese and Chinese-inspired goods as at best foolish trifles worthy of no more serious regard than this season's favoured hat or velvet glove, or at worst as emblems of aesthetic monstrosity or perversion tending only to vitiate the taste of their admirers" (23).

9. Porter, *Chinese Taste*, 23.

10. Porter, *Chinese Taste*, 60.

11. Paul Sandby, *The Analyst Besh[itte]n in his Own Taste*, 1753, etching, 26.4 × 183 cm, British Museum, London.

12. James Wilson, "Essays on the Origin and Natural History of Domestic Animals," *Quarterly Journal of Agriculture* 1 (May 1828–August 1829), 540, 683.

13. Wilhelmina Swainston-Goodger, *The Pug-Dog: Its History and Origin* (London: Watmoughs, 1930).

14. Okumura Masanobu, *Winter: Pictures of the Four Seasons: Plum Blossoms in the Snow*, 1720s, woodblock print, Honolulu Museum of Art, Honolulu.

15. Zhou Fang, *Ladies Wearing Flowers in Their Hair*, late eighth century, ink and color on silk handscroll, 46 × 180 cm, Liaoning Provincial Museum, Shenyang. *Woman Holding a Pekingese*, Tang dynasty (618–907), colored terracotta, 49.2 cm high, Kyoto National Museum, Kyoto. *Figure of a Seated Court Lady*, eighth century, earthenware with sancai glaze, 37.5 × 14.3 × 15.4 cm, Metropolitan Museum of Art, New York. *Chinese Court Lady with Dog*, Tang dynasty (618–907), terracotta, 52 cm high, Timeline Auctions, London (January 14, 2017).

16. "Although there is pictorial, written and zoo-archeological evidence for toy dogs spanning at least the last 2,000 years, no toy breeds possess a basal signature, probably a result of the ease with which they can be transported and interbred with local dogs." Greger Larson, Elinor K. Karlsson, Angela Perri, et al., "Rethinking Dog Domestication by Integrating Genetics, Archeology, and Biogeography," *Proceedings of the National Academy of Sciences of the United States of America*, 109, no. 23 (2012): 8882.

17. Bridgett M. vonHoldt, John P. Pollinger, Kirk E. Lohmueller, et al., "Genome-wide SNP and Haplotype Analyses Reveal a Rich History Underlying Dog Domestication," *Nature* 464 (2010), supplemental material.

18. "Though most breeds within a clade appear to be the result of descent from a common ancestor, the New World dogs and the Asian toys showed nearly 200% improvement in the maximum likelihood score by allowing for admixture between the breeds within the clade. Based on this analysis, the Asian toy dogs were likely not considered separate breeds when first exported from their country of origin resulting in multiple admixture events." Heidi G. Parker, Dayna L. Dreger, Maud Rimbault, et al., "Genomic Analyses Reveal the Influence of Geographic Origin, Migration, and Hybridization on Modern Dog Breed Development," *Cell Reports* 19, no. 4 (2017), 705.

19. Georges-Louis Leclerc, *Histoire Naturelle par Buffon: Quadrupedes* (Paris: Didot, 1799), 1:331.

20. *London Gazette*, December 3–7, 1696; June 6–10, 1706.

21. Hester Lynch Piozzi, *Observations and Reflections Made in the Course of a Journey through France, Italy, and Germany* (London: A. Strahan and T. Cadell, 1789), 1:148.

22. Eliza Haywood, *The Invisible Spy* (London: T. Gardner, 1755), 2:33.

23. Richard Brookes, *The Natural History of Quadrupedes, Including Amphibious Animals, Frogs, and Lizards: With Their Properties and Uses in Medicine* (London: J. Newbery, 1763), 1:225.

24. Yang, "Culture in Miniature," 148–149.

25. Donna Haraway, *The Companion Species Manifesto: Dogs, People, and Significant Otherness* (Chicago: Prickly Paradigm, 2003), 5.

26. Haraway, *Companion Species Manifesto*, 9.

27. Nöel Adams, "Between Myth and Reality: Hunter and Prey in Early Anglo-Saxon Art," in *Representing Beasts in Early Medieval England and Scandinavia*, ed. Michael D. J. Bintley and Thomas J. T. Williams (Woodbridge: Boydell and Brewer, 2015), 20–21.

28. Jacob Coeman, *Pieter Cnoll, Cornelia van Nijenrode and Their Daughters*, 1665, oil on canvas, 132 × 190.5 cm, Rijksmuseum, Amsterdam.

29. Gilbert White, *The Natural History and Antiquities of Selborne, in the County of Southampton* (London: B. White, 1789), 279.

30. John Allan, "Some Post-Medieval Documentary Evidence for the Trade in Ceramics," in *Ceramics and Trade: The Production and Distribution of Later Medieval Pottery in North-West Europe*, ed. Peter Davey and Richard Hodges (Sheffield: University of Sheffield Department of Prehistory and Archaeology, 1983), 42–44.

31. Report, Allen Catchpole to the East India Company Court of Directors, February 10, 1703, India Office Records, IOR/G/12/6, 910, British Library, London. I would like to thank Richard Coulton for bringing this report to my attention.

32. John Bell, *Travels from St. Petersburgh in Russia to Various Parts of Asia* (Edinburgh: William Creech, 1788), 2:17.

33. *A Pair of Paintings of Chinese Women Accompanied by Pipes and Pets*, eighteenth century, gouache on paper, 80 × 59 cm, Martyn Gregory Gallery, London.

34. Olfert Dapper, "A Third Embassy to the Emperor of China and East-Tartary, Under the Conduct of the Lord Pieter van Hoorn, Containing Several Remarks in Their Journey through the Provinces of Fokien, Chekiang, Xantung, and Nanking, to the Imperial Court at Peking," in *Atlas Chinensis: Being a second part of a relation of remarkable passages in two embassies from the East-India Company of the United Provinces . . . to Konchi, Emperor of China and East-Tartary*, trans. John Ogilby (London: John Ogilby, 1671), 240.

35. Dapper, "Third Embassy," 369.

36. Dapper, "Third Embassy," 718.

37. *A Pair of Miniature Oval Portraits*, eighteenth century, oil and gold paint on reverse of mirror, 8.89 × 11.43 cm, Martyn Gregory Gallery, London. *A George III Period Chinese Export Reverse Mirror Painting*, c. 1775, paint on reverse of mirror, 96.5 × 60.5 cm, Ronald Phillips, London.

38. Olof Toreen, "A Voyage to Suratte," a short narrative included in Peter Osbeck, *A Voyage to China and the East Indies*, trans. John Reinhold Forster (London: Benjamin White, 1771), 2:254.

39. Athanasius Kircher, *China monumentis, qua sacris qua profanis, nec non variis naturae & artis spectaculis, aliarumque rerum memorabilium argumentis illustrata* (Amsterdam: Jacob van Meurs, 1667), 115; hereafter abbreviated as *China Illustrata*.

40. Athanasius Kircher, "Some Special Remarks Taken out of Athanasius Kircher's Antiquities of China," in *An Embassy from the East-India Company of the United Provinces, to the Grand Tartar Cham, Emperor of China*, by Johan Nieuhof, ed. and trans. John Ogilby (London: John Ogilby, 1673), 389.

41. Pehr Osbeck, *Dagbok öfwer en Ostindisk resa åren 1750, 1751, 1752.* (Stockholm: Lor. Ludv. Grefing, 1757). Peter Osbeck, *Reise nach Ostindien und China*, ed. and trans. Johann Gottlieb Georgi (Rostock: Johann Christian Koppe, 1765).

42. Paulus Potter, *A Hall Interior with a Group of Eight Dogs, Two Seated upon a Green Cushion, the Others Resting and Playing on the Ground*, 1649, oil on canvas, 119 × 150 cm, Noortman Master Paintings, Amsterdam. Hubert Gerhard, *Female Pug*, c. 1600, cast bronze, 21 cm high, Museumslandschaft Hessen Kassel, Kassel. August Erich, *Landgrave Maurice of Hesse-Kassel with his Family*, 1618–1628,

oil on canvas, 230.5 × 422 cm, Museumslandschaft Hessen Kassel, Kassel. Hans Rottenhammer, *Minerva with the Muses on Mount Helicon*, 1603, oil on canvas, 186 × 308.8 cm, Germanisches National-museum, Nuremberg. All three animals represent very early stages in the development of the breed and bear little resemblance to modern-day pugs. The appearance of the pug became vaguely standardized only in the eighteenth century; stricter codification began much later with the development of dog breeding for show in the Victorian period.

43. Hendrick Goltzius, *Lot and His Daughters*, 1616, oil on canvas, 140 × 204 cm, Rijksmuseum, Amsterdam. Hendrick Goltzius, *Justus Lipsius*, 1587, engraving, 14 × 9.5 cm, National Gallery of Art, Washington, DC. "Mopshond" or "Mops" eventually became the Dutch word for "pug."

44. Giovanna Garzoni, *Bitch with Biscuits and Chinese Bowl*, c. 1648, oil on canvas, 27.5 × 39.5 cm, Pitti Palace, Florence. Giovanna Garzoni, *Pug on a Table with Bread, Chinese Tea Bowl, Sopressata and Flies*, mid-seventeenth century, watercolor on vellum, 27.3 × 36.8 cm, Doyle New York, New York (January 26, 2015).

45. Jacques Roger, *Buffon: A Life in Natural History*, ed. L. Pearce Williams, trans. Sarah Lucille Bonnefoi (Ithaca, NY: Cornell University Press, 1997), 217.

46. Francis Coventry, *The History of Pompey the Little: Or, the Life and Adventures of a Lap-Dog* (London: M. Cooper, 1751).

47. Ingrid H. Tague, *Animal Companions: Pets and Social Change in Eighteenth-Century Britain* (University Park: Pennsylvania State University Press, 2015), 22.

48. Horace Walpole to the Hon. Thomas Walpole, Strawberry Hill, October 26, 1780, in *The Yale Edition of Horace Walpole's Correspondence*, ed. W. S. Lewis, vol. 36, *Horace Walpole's Correspondence with the Walpole Family*, ed. W. S. Lewis and Joseph W. Reed Jr. (New Haven, CT: Yale University Press, 1973), 181–182.

49. David Beevers, "'Mand'rin only is the man of taste': 17th and 18th Century Chinoiserie in Britain," in *Chinese Whispers: Chinoiserie in Britain, 1650–1930*, ed. David Beevers (Brighton: Royal Pavilion and Museums, 2008), 19.

50. "Upon a new Marry'd Lady's being fond of a *Dutch Mastiff,* which caus'd her Husband to be Jealous," *Diverting Post* 19 (February 24— March 3, 1705), 1.

51. Beevers, "Mand'rin only," 21, notes that while oriental consumer goods and chinoiserie were generally "closely connected with notions of femininity and female patronage . . . japanning, the imitation of oriental lacquer, was considered to be a particularly female accomplishment."

52. Richard Graves, "The Insolence of Office," in *Euphrosyne: Or, the Amusements of the Road of Life* (London: James Dodsley, 1776), 88.

53. Alexander Pope, *The Rape of the Lock, An Heroi-Comical Poem in Five Canto's,* 4th ed. (London: Bernard Lintott, 1715), 29.

54. The Pope quotation opens Elizabeth Kowaleski-Wallace's essay "Women, China, and Consumer Culture in Eighteenth-Century England," in *Eighteenth-Century Studies* 29, no. 2 (1995), 153.

55. Charles Molloy, *The Coquet: Or, the English Chevalier* (London: E. Curll and R. Francklin, 1718), 31. For Yang on Molloy, see "Culture in Miniature," 154.

56. Tague, *Animal Companions,* 108.

57. Margery Daw (pseud.), "Literary Lady," *British Stage and Literary Cabinet* 5, no. 59 (1821), 381.

58. Robert Dodsley, *The Toy-Shop. To which are added, Poems and Epistles on Several Occasions,* 11th ed. (London: Robert Dodsley, 1744), 15.

59. Haywood, *Invisible Spy,* 2:34.

60. Susan Ferrier, *Marriage,* ed. Herbert Foltinek (Oxford: Oxford University Press, 2001), 132–133.

61. John P. Cushion, *Animals in Pottery and Porcelain* (London: Studio Vista, 1974), 173. The form of these Chinese ceramic dogs changed little over the course of centuries; an earthenware Tang dynasty (618–907) hound bears a considerable resemblance to two porcelain dogs made in Dehua and Jingdezhen in the eighteenth century. *Figure,* Tang dynasty (618–907), unglazed earthenware, 12 × 4 × 9 cm, FE.155–1974, Victoria and Albert Museum, London (hereafter V&A). *Figure of a Dog,* c. 1700–1750, porcelain with a white glaze, 15.2 cm × 14 cm, 4823–1901, V&A. *Figure of a Dog,* eighteenth century, porcelain painted with overglaze enamels and gilt paint, 18.5 × 11.3 cm, FE.11A–1978, V&A.

62. Johann Joachim Kändler, *Die Arbeitsberichte des Meissener Porzellan-modelleurs Johann Joachim Kaendler 1706–1775*, ed. Ulrich Pietsch (Leipzig: Edition Leipzig, 2002), 23–24.

63. Dennis G. Rice, *Dogs in English Porcelain of the 19th Century* (Woodbridge: Antique Collectors' Club, 2002), 69.

64. Gabriel-Louis Calabre Pérau, *L'Ordre de Francs-Maçons Trahi et le Secret des Mopses révélé* (Amsterdam: Henri-Albert Gross & Co., 1745).

65. Wolfgang Saal and Frithjof Schwartz, *Der Mops: Ein Kunstwerk*, trans. Judith Rosenhal (Trier: H2SF, 2008), 25–29. Ute Christina Koch, "Count Brühl and His Collection of Porcelain Boxes," in *Going for Gold: Craftsmanship and Collecting of Gold Boxes*, ed. Tessa Murdoch and Heike Zech (Brighton: Sussex Academic Press, 2014), 189–190.

66. Patricia F. Ferguson, "Sprimont's Complaint: Buying and Selling Continental Porcelain in London (1730–1753)," in *Arts Antiques London Catalogue: 2012* (London: Haughton International Fairs, 2012), 13.

67. Harry Phillips, *The Valuable Library of Books in Fonthill Abbey: A Catalogue of the Magnificent, Rare and Valuable Library (of 20,000 Volumes)* (London: Phillips, 1823), 2:187.

68. Johann Joachim Kändler (Meissen Porcelain Manufactory), *Pug Dog*, c. 1741–1745, porcelain with enamel paint, 17.3 × 19.2 × 11.1 cm, PE 3895, Porzellansammlung, Staatliche Kunstsammlungen Dresden. Saint-Cloud Porcelain Manufactory, *Figure of a Pug-Dog*, mid-eighteenth century, porcelain painted in enamels, 20 × 23.5 cm, C.317–1909, V&A. Vincennes Porcelain Manufactory, *Pair of Pug Dogs*, c. 1750–1752, soft-paste porcelain on mount, 20 × 19 cm, Artcurial, Paris (June 20, 2006). Lowestoft Porcelain Factory, *Pug*, c. 1755–1760, porcelain with manganese coloring, 9.21 cm high, 1988.985, Boston Museum of Fine Arts, Boston. Longton Hall Factory, *Pair of Pugs Seated on Their Haunches with Curled Tails*, c. 1753–1757, soft-paste porcelain with underglaze manganese decoration, 8.9 cm and 9.2 cm high, Mr. and Mrs. Sigmund Katz Collection, Boston Museum of Fine Arts. William Duesbury and Co., *Pug Dog*, 1760, soft-paste porcelain painted in enamels, 8.3 cm high, 414.209/A–1885, V&A.

69. *A Pair of Erfurt Pug Tureens and Covers*, c. 1760, manganese, 16 cm wide, Christie's, London (July 8, 2002). *Bonbonnière*, c. 1770, faience, 14 × 25 cm, private collection. *Belgian Faience Polychrome Model of a Pug*,

late eighteenth century, faience, 8.9 cm high, Christie's, London (September 16, 2012). *Pug on a Cushion*, c. 1745–1755, white salt-glazed stoneware, 4×5 cm, C. 800–1928, Fitzwilliam Museum, Cambridge. *Bonbonnière*, 1760–1780, painted enamel, 5.7×6.3 cm, B1644, Black Country Museums, Wolverhampton.

70. François Boucher, *A Young Lady Holding a Pug Dog*, c. 1745, oil on canvas, 34.5×28.6 cm, Art Gallery of New South Wales, Sydney. Henri Welschinger, "Le chien du duc d'Enghien," *Le Monde Illustré* 1656 (December 22, 1888), 403–406.

71. David Howard and John Ayers, *China for the West: Chinese Porcelain and Other Decorative Arts for Export Illustrated from the Mottahedeh Collection* (London: Sotheby Parke Bernet, 1978), 2:596.

72. *A Rare Pair of Chinese Export Figures of Seated Pug Dogs*.

73. Howard and Ayers, *China for the West*, 600.

74. See, for instance, "A Chinese Export Seated Pug Dog," Christie's Auctions and Private Sales. http://www.christies.com/lotfinder/Lot/a -chinese-export-seated-pug-dog-qianlong-6033487-details.aspx; *A Chinese Export Seated Pug Dog*, Qianlong period (1736–1795), porcelain, 24.5 cm high, Christie's, London (sale 12291, November 9, 2016; lot 491). In contrast to the Meissen pugs described at the beginning of this essay, this pug does not have a fake Meissen mark. It is also posed differently and painted in a different style and color.

75. Yang, "Culture in Miniature," 167.

76. *A Meissen Plate with the "Parrot and Spaniel" Pattern*, c. 1740, hard-paste porcelain, 23.6 cm diameter, Bonhams, London (December 7, 2011).

77. A cup and a saucer of this set survive in the collections of the V&A in London: *Cup and Saucer*, c. 1750, porcelain, cup 3.8 cm high, 7.9 cm diameter; saucer 12.4 cm diameter, 642&A–1903, V&A.

78. Johann Gregorius Höroldt (Meissen Porcelain Manufactory), *Meissen Chinoiserie Rinsing Bowl*, 1723–1724, hard-paste porcelain, 8.5×17.7 cm, CE.74.131, Smithsonian National Museum of American History, Washington, DC. Johann Gregorius Höroldt (Meissen Porcelain Manufactory), *Tankard*, c. 1725, hard-paste porcelain decorated in underglaze blue, 19.4×15.1×10.2 cm, 1980.615, Museum of Fine Arts, Boston.

79. Surviving examples include a *Figure*, early eighteenth century, blanc de chine porcelain, 14.3×15.55 cm, C.108–1963, V&A; and *Man and Woman*, eighteenth century, porcelain decorated with enamels and gilding, 16.5×15 cm, C12–1951, V&A.

80. Bow Porcelain Factory, *Mug and Cover*, c. 1755–1760, soft-paste porcelain, 20.9 cm high, 12.1 cm diameter, 414.109&/A–1885, V&A.

3

Green Rubies from the Ganges

Eighteenth-Century Gardening as Intercultural Networking

BÄRBEL CZENNIA

Thou by the Indian Ganges' side
Should'st rubies find:
—Andrew Marvell

A network most broadly defined is a structure made from various types of interlacing material. The term is also used more metaphorically for complex patterns of communicating objects as in road systems or communication systems. Last but not least, it may describe groups of interconnected people "who exchange information, contacts, and experience for professional or social purposes," as in the term "trade network" or "support network."[1] All three aspects of networks come into play in the creation of gardens. Every garden is quite literally a network: Its vegetal components are physically linked or interwoven above and below ground by interlacing branches and root systems. A garden is also an ecological network with complex "relationships among organisms, including people, and the non-living world."[2] Moreover, a garden as a premeditated aesthetic design relies on a network of corresponding horizontal, vertical, and diagonal reference lines, translated into materials that speak to the human senses: visual patterns of color and texture made of flowers, leaves, bark, soil, and stone; patterns of sound generated by rustling vegetation, sprinkling water, and resident or visiting animals; patterns of fragrances emanating from flowers; even tactile patterns experienced underfoot such as gravel, sand, wood, pavement, or lawn. Many gardens also owe their existence to a multitude of interpersonal and institutional networks formed outside their enclosures. This essay examines the influence of various types of networks on the exchange of plants, gardeners, garden concepts, and garden architecture between Britain and British India during the long eighteenth century. It explores the effect of cultural and commercial networks as well as of human networking on the design of green spaces in East and West. By analyzing cultural and material dimensions of "human-environmental interactions," recently discussed by James Beattie and colleagues as "eco-cultural networks,"[3] this essay also applies new perspectives developed by imperial historiography to a case study that links garden history and art history with literary history.

Plant Networks and Human Networks

Gardens are highly eclectic environments, populated with species that could never have shared habitats without human interference. Vasco da Gama's first successful journey to India resulted in a significant uptake of tropical plant imports to Europe along the first maritime link between West and East.[4] During the sixteenth century, informal networks of European plant collectors, including sailors, soldiers, medical doctors, apothecaries, gardeners, and botanists began to comb South Asia in search of greens for medicinal, economic, or ornamental purposes.[5] Early modern travel reports abound with references to the indigenous vegetation and gardens of the Indian subcontinent and to their exotic appeal for European visitors. François Bernier, who visited northern India in the 1660s, called Kashmir "the Terrestrial Paradise of the Indies" and concluded: "The whole kingdom wears the appearance of a fertile and highly cultivated garden."[6] Domingo Fernández de Navarrete, who traveled from Madras to Golconda in 1670, enthused over "infinite groves of wild palm-trees," "woods . . . of tamarine-trees," "stately orchards" planted with "innumerable oranges and lemons," and palaces surrounded by "walks . . . wide and very clean, with ponds at distances, and water-works continually playing . . . fit for any prince."[7] Dean Mahomet, the first native-born travel writer to describe the beauties of India to British audiences in the English language, remembered his former homeland as a Miltonic paradise:

> The people of India . . . are peculiarly favoured by Providence . . . and tho' the situation of Eden is only traced in the Poet's creative fancy, the traveller beholds with admiration the face of this delightful country, on which he discovers tracts that resemble those so finely drawn by the animated pencil of Milton. You will here behold the generous soil crowned with various plenty; the gardens beautifully diversified with the gayest flowers diffusing their fragrance on the bosom of the air.[8]

Describing the region of Alahabad shortly after the end of the rainy season, he added:

> The gardens are painted with a variety of beautiful flowers . . . to the rose, and a white flower resembling jessamine, we are only indebted for their fragrance. The fruits are mangoes, guavas, pomegranates, ananas or pine apples, musk and water melons, limes, lemons, and oranges, all which spring up spontaneously, and grow to a great degree of perfection. Ginger, and turmeric . . . are produced . . . in their highest state of excellence.[9]

The image of India as an ever-flourishing garden appealed to British empire builders as much as to poets based in a northern climate where vegetation cycles were regularly interrupted by harsh winters.[10] Before and after Milton employed the "figtree . . . such as at this day to Indians known / In Malabar or Deccan"[11] to help Adam and Eve cover up after the fall, English readers relished accounts of the sprawling banyan tree. In 1717, Thomas Tickell even compared Princess Caroline, wife of future George II, to "the fam'd Banian tree, whose pliant shoot / To earthward bending of itself takes root, / Till like their mother plant ten thousand stand / In verdant arches on the fertile land," suggesting that her numerous offspring would in due course spread Hanoverian genes and influence to royal houses all over Europe.[12] Robert Southey's epic *The Curse of Kehama* dedicated several stanzas to "an aged banian" forming the center point of a Hindu village.[13] Visual representations of banyan trees accessible to eighteenth-century viewers included William Hodges's *Natives Drawing Water from a Pond with Warren Hastings' House at Alipur in the Distance* (1781),[14] Auguste Borget's *Moonlit Scene of Indian Figures and Elephants among Banyan Trees, Upper India (probably Lucknow)* (c. 1787),[15] and Benjamin Thomas Pouncy's engraving *Banyan Tree after William Hodges, 1793*.[16] Thomas and William Daniell's popular collection *Oriental Scenery* (1795–1816)[17] and James Forbes's *Oriental Memoirs* (1813)[18] also offered folio-sized representations that familiarized British audiences with this Asian plant.

During a period of fiercely competitive nation building, however, the lush vegetation of the East was often presented as inseparable from arbitrary government.[19] James Thomson's poem *Summer* (1727) linked Asia's "gay profusion," "pomp of Nature," "balmy meads," "powerful herbs," "ambrosial foods, rich gums, and spicy health," and "silky pride" to a coexisting "world of slaves" and an "oppressive ray the roseate bloom / Of beauty blasting." The less spectacular vegetation of "milder climes" was redeemed by "the softening arts of peace," "Progressive truth," "Kind equal rule," "government of laws," and "all-protecting freedom."[20] Just as Thomson's *Liberty* (1734) counterbalanced "the gorgeous east" and "Asia's Woods," "vegetable fleece," and "more delicious fruits" with Britannia's "freedom,"[21] so did Alexander Pope's *Windsor-Forest* (1713) contrast India's opulent vegetation with a "barb'rous Ganges" armed by a "servile Train."[22]

But while patriotic poets liked to pit English liberty against Eastern tyranny, political polemics never seriously interfered with British enthusiasm for "citron groves," the "spreading tamarind" and "its fever-cooling fruit," "massy locust sheds," "the Indian fig," "verdant cedar wave / And high palmettos," "the cocoa's milky bowl," "the palm" and its "freshening wine," or "the full pomegranate" in Eastern gardens.[23] The plants that had stimulated the imagination of Milton's, Thomson's, and Pope's readers during the first half of the eighteenth century continued to fascinate armchair travelers lured to the Orient in the last decade of the century by Dean Mahomet. Britain's growing political influence in India was also reflected in the rising number of tropical plants listed and the descriptive detail offered in scientific and nonscientific publications of the period. As familiarity with Indian plants increased, so did the desire of English gardeners to make them part of their own world.

James Forbes's descriptions of Indian gardens in the early nineteenth century, for example, introduced countless other plants beside the "banian" or "burr tree," including the cocoa nut tree, the palmyra or "brab" tree, date palm, mango tree, custard apple, "champach" tree (i.e., *Michelia champaca* or *Magnolia champaca*), "keurah" tree (also known as kewra or pandanus), the rose-acacia or "bawbul" tree (also known as "babul," a member of the mimosa family), the "jaca" tree (i.e., Jack tree), tamarind, and the "cajew" or "cashew apple." He also praised the "mogree" (or mogra, a type of jasmine), oleander, and myrtle, various Indian cereals and vegetables, including "banda" (*Hibiscus esculentus*), "bungal," yam, fenugreek, Indian beans, chili pepper, cadamon, and ginger. Adding to his detailed verbal accounts, Forbes also showcased decorative Indian flowers such as lotus, (pink) oil plant, "tube-rose," and "hinna" with his own colorized illustrations.

Even Alexander Pope's earlier, seemingly depreciative comment about Indian vegetation in *Windsor-Forest* must be taken with a grain of salt. More political rhetoric (in the contemporary debate about the ending of the Spanish War of Succession) than horticultural advice, the lines, "Let *India* boast her Plants, nor envy we /

The weeping Amber or the balmy Tree, / While by our Oaks the precious Loads are born, / And Realms commanded which those Trees adorn,"[24] hardly intended to discourage English interest in exotic plants. Nor did they keep Pope the hobby gardener from planting Asian species himself. As soon as the "oaks of Windsor" hit the water in the service of Britain's "peaceful" (l. 364) new trade empire, they offered stowage space not only for "amber," "coral," "ruby," and "pearly shell" (ll. 391–393) but also for new seeds, bulbs, and flower pots, filled with novelty samples of Indian vegetation.

References to specific plants in Twickenham suggest that Pope adorned his own garden with species that were still regarded as exotic and "foreign" during his lifetime. The most obvious example may have been a "Weeping Willow," falsely named *Salix babylonica* by Linnaeus but originating in northern China.[25] "An Epistolary Description of the Late Mr. Pope's House and Gardens at Twickenham," first published anonymously in 1748, mentions "lawrel and bay" (240), plants that originate in the Mediterranean region.[26] In "The Second Satire of the Second Book of Horace Paraphrased" (1733–1734), Pope himself alludes to "grapes," "figs," a "wallnut-tree," and "Broccoli."[27] Broccoli had a long history of cultivation in Italy but was still a novelty in eighteenth-century Britain, where it was introduced via Antwerp. Pope's "grapes," "figs," and "walnut" remind us of the relativity of plant designations as "foreign" or "indigenous." While Paul Baines has interpreted Pope's lines as advocating a withdrawal from the "world of empire" and a return to the "ancient, maturing fruits of the English garden,"[28] it could be argued that even plants that reached England sometime between Roman occupation and the Renaissance period were, after all, not strictly endemic or "English" but originally introduced from the Near East (grapes), the Middle East (figs such as Ficus carica), or even the Far East (walnut trees such as *Juglans regia*, despite being called "Old World Walnut" or "English Walnut").[29] As early as 1711, Joseph Addison had contested the idea of genuinely "English" plants and celebrated transregional exchange ("Traffick") through networks of commerce as a means to overcome England's natural lack of botanical diversity:

If we consider our own Country in its natural Prospect, without any of the Benefits and Advantages of Commerce, what a barren uncomfortable Spot of Earth falls to our Share! Natural Historians tell us, that no Fruit grows originally among us, besides Hips and Haws, Acorns and Pig-Nutts, with other Delicacies of the like Nature; That our Climate of it self, and without the Assistances of Art, can make no further Advances towards a Plumb than to a Sloe, and carries an Apple to no greater a Perfection than a Crab: That our Melons, our Peaches, our Figs, our Apricots, and Cherries, are Strangers among us, imported in different Ages, and naturalized in our *English* gardens. . . . Nor has Traffick more enriched our Vegetable World, than it has improved the whole Face of Nature among us. Our Ships are laden with the Harvest of every Climate. . . . Nature indeed furnishes us with the bare Necessaries of Life, but Traffick gives us a great Variety of what is Useful, and at the same time supplies us with every thing that is Convenient and Ornamental.[30]

Ever since the Glorious Revolution, when William of Orange "introduced Dutch glass house technology and a taste for exotics to Britain," the number of transplants from the "East Indies" raised in private or public hothouses and listed in garden inventories or garden manuals grew continuously.[31] Philip Miller, chief gardener at the Chelsea Physic Garden from 1722 to 1771 and member of a "plant-collecting and botanical network that stretched to every corner of the globe,"[32] described Indian novelties such as "Corindum," "Yellow Indian Jasmine," and the "Indian Lily-Daffodil." His *Gardeners Dictionary* (1731) also provided detailed building and maintenance instructions for stove houses or "stoves," suitable for "Cashew," "Custard Apple," "the Arched Indian . . . Fig Tree," "Ginger," and "Tamarind Tree."[33] Over the following fifty years, the number of references to plant imports from India grew continuously. In the first edition of *Hortus Kewensis*, William Aiton, former assistant to Philip Miller in Chelsea and first director of the botanic garden in Kew (established in 1759) described thirty-five grasses, herbs, flowers, shrubs, and trees of Indian origin, successfully cultivated by 1789.[34] The list included "Indian sage," "Bamboo Reed-grass," "Indian Plantain," "Indian Oak, or Teakwood,"[35] "Common Sweet Basil," "Spotted-bark'd Cotton-Tree,"[36] "Scarlet Kidney-Bean," "Dyer's Indigo," and the "Shaddock-tree," a citrus tree variously referenced as an early version of grapefruit or pomelo.[37] To these numbers could be added several "stove plants" from the "East Indies," including the "Dragon Tree," the "Flowering Indian Gardenia," the "Asiatic Paulinia," "Asiatic shrubby Psychotria," and the "Indian Tamarind Tree," which John Abercrombie had recommended to wealthy English garden owners five years earlier.[38]

Freshly returned to England for impeachment and a new life as gentleman farmer on his Daylesford estate, Gloucestershire (purchased in 1788), Warren Hastings, former governor general of Bengal and first de facto governor general of India, asked his Indian agents "to send him lychee, cinnamon, and custard apple seeds . . . so that he could recreate" his Indian garden on British soil. His addition of "an orangery heated with hot water pipes"[39] to more traditional "pleasure gardens," "lakes," and a "well-timbered park" illustrates how the transculturated taste of British world citizens affected eighteenth-century English country estates. Overlapping personal and professional networks of merchants, military personnel, and imperial administrators, especially those who had spent decades in the service of the East India Company, significantly increased the biological diversity of Britain's gardens. As Ray Desmond has shown, Hastings was neither the first nor the last returning nabob to do so:

Charles du Bois (1656–1740) used his position as the Company's Treasurer to procure plants from India for his garden at Mitcham in Surrey. Sir Joseph Banks in return for services rendered to the Court of Directors was able to sow at Spring Grove seeds received from Koenig, Kyd, Roxburgh, and Sir John Murray. . . .

The real impact of the Indian subcontinent lies in its contributions to the English garden flora."[40]

While Hastings grew Indian plants in Gloucestershire, homesick company men in Bengal inversely yearned for flowers that they associated with their earlier lives in Britain. A 1794 advertisement by the "Europe, China, and India Warehouse" published in the *Calcutta Gazette* alerted expats to a fresh load of garden flowers[41] newly arrived from England, and, even better, tested and proven to germinate in Indian soil, despite the hazardous sea journey.[42] According to Eugenia Herbert, imports included flowers popular in English cottage gardens such as "stock, carnation, marigold,[43] larkspur, mignonette, honeysuckle, polyanthus,[44] poppy and Sweet William." Many of these European transplants would embellish the secondary residences or "garden homes" built along the river Hooghly on the outskirts of the city to provide temporary escape from Calcutta's summer "heat, filth, and plagues."[45] How urgent the craving of expats for "English" flowers could become, is obvious from a letter by missionary and hobby gardener William Carey to his brother Sutcliff in 1809:

Were you to give a penny a day to a boy to gather seeds of cowslips, violets, daisies, crowfoots &c., and to dig up the roots of bluebells, &c., after they have done flowering, you might fill me a box every quarter of a year; and surely some neighbours would send a few snow-drops, crocuses, &c., and other trifles. All your weeds, even your nettles and thistles, are taken the greatest care of by me here.[46]

While personal networks of garden lovers abroad and at home were very important for Anglo-Indian horticultural exchanges, the shipment of Indian plants and seeds was not limited to private initiatives. Professional naturalists, botanists, and gardeners also collaborated within and across a variety of new, publicly funded or government-operated networks.

Most significant for plant exchanges between Britain and its Eastern trading posts was an emerging institutional network of professionally managed botanic and cultivational gardens, linked by a scientific and geopolitical agenda, close communication ties, and a high degree of personnel interchange.[47] The transformation of university-owned physic gardens (with a special interest in medicinal plants, for instance, in Oxford and Edinburgh) and royal estates (with a special interest in decorative exotic plants, for instance, in Richmond and Kew) into botanic gardens with research programs, specialty collections, or clearly defined service functions within the imperial economy could probably be rewritten as a history of overlapping networks. One important mediator figure between already established and emerging new scientific networks was Joseph Banks. A botanist himself and longtime president of the Royal Society of London, a research institution with its own global network of corresponding members, Banks also served as longtime informal adviser to George III on the botanic garden in Kew, opened in 1759. Under the unofficial leadership of Banks, Kew became the hub of a worldwide network of botanical knowledge. Its nodes combined scientific exploration with systematic relocation and redistribution of plants that promised to give the British Empire a competitive edge over its international rivals in an increasingly globalized economy.[48] Plants of medicinal and ornamental value as well as potential new cash crops (such as teak, spices, cotton, indigo, tea, hemp, gum trees, and many others) were tested in these research and distribution centers before being renaturalized in other parts of the British Empire.

The Royal Botanic Garden of Calcutta, for instance, founded by Colonel Robert Kyd in 1787 as the first of thirty-four such gardens created on the Indian subcontinent during British rule,[49] facilitated the translocation of Asian cash crops, including *Camellia sinensis* from China, to more westerly parts of the British Empire. Fostering transregional connectivity, the Calcutta garden also served as Eastern hub for the shipping of botanic transplants to the new colony of Australia and to Europe, with the botanic garden of the Cape Colony serving as intermediary node.[50] A global network of botanic gardens, one might say in variation of a sentence by Andrea Wulf, "not only changed the English landscape but the very fabric of the nation, contributing to the country's global dominance and imperial strength."[51]

Not only did Joseph Banks alert "Farmer George" to the importance of botanic gardens for empire building; Banks also used his friendship with the king and with many botanists and plant collectors around the world to influence the filling of leadership positions (such as superintendents and head gardeners) in newly founded botanic gardens, to shape collection programs, encourage exchanges, and advise local staff. Banks strengthened the emerging institutional network of government-coordinated botanical research centers with additional layers of personal networks that linked Kew-trained gardeners and other specialists affiliated with the Royal Society in a global network of garden professionals.[52] Banks also organized expeditions to collect plants in many parts of the world and used his personal ties with botanists and sea captains so successfully that by the early nineteenth century many British ships returning from imperial outposts carried foreign plant samples destined for Kew. Contemporary reports by eyewitnesses suggest that, upon arrival, some of the vessels appeared literally transformed into floating gardens.[53]

Each node within the new global network of botanic gardens was in turn locally linked to other formal or informal institutions. In the Eastern Hemisphere, learned societies with an orientalist focus included the Asiatic Society of Bengal, founded in 1784 by Sir William Jones in Calcutta, and the Tranquebarian Society, founded in 1788 by Henning Munch Engelhart in the Danish-Norwegian colony of Tranquebar on the Coromandel Coast.[54] As local "nodes of science"[55] within a global network of knowledge, both institutions maintained their own libraries, museums, and local botanic gardens as well as their own journals, which could be regarded as nodes of a (nested) network of worldwide scholarly communication. Modeled on the *Transactions of the Royal Society of London*,[56] the *Asiatick Researches: or, Transactions of the Society, Instituted in Bengal, for Inquiring into the History and Antiquities, the Arts, Sciences, and Literature, of Asia* and the "Transactions of the Tranquebar Society" (*Det Tranquebarske Selskabs Skrifter*) provided publication outlets and discussion forums for an international community of botanists exploring the Indian flora.

Although this (communication) network within a larger network of science primarily targeted specialist authors and readers, articles on Asian plants in both journals were also accessible to educated but nonprofessional gardeners around the world at a time of fluid boundaries between amateur and professional scientists. Linking specialist and nonspecialist audiences with an interest in Asian flora, the journals contributed to a growing interest in plants from India among British hobby gardeners. Their constant demand for new types of tropical plants was met by professional nurseries, including many that were based in London and "expanded spectacularly from about the mid-eighteenth century."[57] The specialist nurseries were often run by former head gardeners of botanic gardens such as Kew, thereby providing personalized links between the emerging institutional garden network and the private Anglo-Indian horticultural networks described in my previous section, "Plant Networks and Human Networks."

The advertising strategies developed by professional nurseries also resulted in an entirely new genre of literature: illustrated periodicals such as William Curtis's *Botanical Magazine* (since 1787) and Conrad Loddiges's *Botanical Cabinet* (since 1817) provided yet another source of knowledge linking amateur gardeners and professionals. By featuring and popularizing new Indian specimen and instructing a nonspecialist readership in the best Enlightenment tradition, they provided a new communication network for an "imagined community" of cosmopolitan plant lovers.[58]

Johan Zoffany's painting *Warren Hastings and His Second Wife in their Garden at Alipore* (c. 1784–1787)[59] bears witness to the global success story that the English landscape garden had become by the later eighteenth century: Capability Brown's sweeping lawns with their seemingly natural tree clumps had arrived in the East. Temporarily editing out the elephant in the far distance and the exotic apparel of a servant figure at Mrs. Hastings's side, viewers might take the mansion in the background to be just another Palladian residence in the British homeland. The same holds true for *Allington Hunting Party in India*, attributed to Thomas or William Daniell.[60] By emphasizing extensive lawns interspersed with groups of assiduous trees and the prospect of a white palace in the far distance, both paintings evoke the characteristic sense of wide open space that William Kent, Charles Bridgeman, and their successors created for their clients in England.

Dean Mahomet had already observed typical features of English landscape gardens in India during his travels in the 1770s. Moving around as camp follower to an Anglo-Irish officer in the army of the East India Company, he took note of "winding walks planted with embowering trees on each side, and fish ponds reflecting, like an extended mirror, their blooming verdure on each margin" near Calcutta (1772); an "avenue . . . shaded on both sides, with rows of embowering trees," "fishponds, cascades, and groves," and "expanding lawns . . . adorned with figures of snow-white marble" in Chandernagore (1778).[61] "If I had not from time to time seen a tall coconut tree towering above the rest, I could have fancied myself on the banks of the Thames," Bishop Heber wrote in 1820 about Barrackpore.[62] Even some members of the Indian ruling elite adopted the English landscape garden to express their political alignment with their mighty British neighbors.[63]

British visual artists who had traveled in India were often connected to wealthy expatriates working for the East India Company or otherwise employed in the service of empire. Tied into networks of personal friendship and business resulting from art commissions, these painters employed various techniques to make their representations of Indian landscapes look more European.[64] Painterly Anglicizations of Indian landscapes heightened the appeal of art works catering to the Western taste of British collectors at home and abroad.

Corresponding attempts at horticultural mimicry on the part of expatriate gentleman gardeners and landscape architects took various forms. One possibility, indirectly documented by Zoffany's painting, was to opt for more viable native trees, but to select species with silhouettes or textures that looked "English" at least from a distance.[65] However, this strategy did not always result in the perfect effects British garden owners in India had hoped for.[66] What paintings in contrast to private diaries and letters did not show or, perhaps, deliberately edited out was that the English landscape garden in India was a cultural hybrid out of biological necessity. English oaks and elms struggled in India's tropical climate zone. Quite literally unable to mesh with the local network of plants, they often had to be replaced by endemic species. In contrast to the tree behind Thomas Gainsborough's *Mr. and Mrs. Andrews* (c. 1750),[67] the sturdy giant behind Zoffany's Mr. and Mrs. Hastings was not an oak but a banyan tree.[68]

Gardeners met with even greater horticultural challenges where they attempted to transplant a much smaller, yet nevertheless aesthetically significant component of English landscape gardens, English grass: "Soon after my arrival at Baroche," Scottish writer and artist James Forbes reported in his *Oriental Memoirs*,

I purchased a small house and some land in the village of Vezel-poor. . . . My garden occupied about six acres; *I formed it as much as possible after the English taste*, and *spared no pains to procure plants and flowers from different parts of India and China*: it contained several large *mango, tamarind, and burr-trees*, which formed a delightful shade; besides a variety of *smaller fruit-trees and flowering-shrubs*. . . . *One great desideratum is the verdant lawn almost peculiar to the English gardens*; a tropical sun would not admit of it in the fair season, and during the rainy months the rank luxuriant grass more resembles reeds and rushes than *the soft carpet bordered by an English shrubbery.*[69]

At a later point in his Indian travels, Forbes lamented that "all the Asiatic gardens I ever saw, were *deficient in the verdant lawns,* artless shrubberies, and varied scenery, which when attempted to be introduced in France, Italy, and Germany, I always found dignified by the appellation of 'Jardins à l'Angloise.'"[70] Forbes's description of irrigation practices at Baroche, involving "three men and a boy eight hours every day,"[71] illustrates the extravagant amount of human and nonhuman resources that was necessary to sustain an "oriental" garden with a transregional plant community. It also highlights the garden's dependence on the availability of cheap labor in an imperial economy.

Gardeners who accepted the futility of a constant fight against the Indian climate discovered that the want of adaptable English plants could be counterbalanced by emphasizing architectural elements associated with the English gardening tradition: "Vines and creeping plants were trained to conceal two pillars of rude construction . . . one of these I entirely covered with the lively ipomea . . . the other I modernized a little in the European taste, and placed an urn on the summit, dedicated to the naiad."[72] Neither of these strategies could conceal the fact that the English landscape garden in India was a transculturated assemblage, with species from many corners of the empire linking forces above and below ground in hybrid ecological networks. Remembering his second landscape garden, at Zulam-Bhaug, Forbes wrote:

Although the pavilions and other buildings were less magnificent, the grounds were more artless and beautiful than the generality of the Indian gardens; and profusely adorned with trees, shrubs, and flowers; not only of those indigenous to Hindostan, but with every variety procurable from China, Persia, and Europe. The apple and the peach, flourishing with the Chinese roses and oranges, interspersed among mangos, plantains, and tamarinds.[73]

Just as Forbes the gardener fused Western and Eastern components in his Anglo-Indian landscape design, Forbes the creative writer blended European neoclassicism with Indian mythology in his garden poetry. In a poem explicitly composed for the pedestal of one of his decorative urns, he transformed the familiar "naiad" of ancient Greece into "Medhumadha, a water nymph in the Hindoo mythology," tracing "balmy zephyrs" back to the breath of "Indra," the Indian "god of the seasons," and replacing Greek Apollo by "Mitra, a solar deity of the Hindoos."[74] Resisting wholesale Europeanization, Indian gardens reminded their British owners and visitors that empire building was not a one-way road but a process that transformed people, poems, gardens as well as the garden aesthetics in both hemispheres. Transculturation, however, was not always perceived as a cultural enhancement. As Tillman Nechtman has observed, "[t]aken as a group, late-eighteenth-century nabobs were cultural threats because they brought empire home and threatened to naturalize it as part of the national landscape" and "exposed the fluid dynamics that moved between empire and nation."[75]

Compared to homesick British expats stationed in India and yearning for "English" flowers, English nabobs returning to their old homeland tended to be more appreciative of the natural and cultural otherness of Asian environments they left behind. Several imperial administrators inscribed their new Anglo-Indian rather than British identity into the gardens they acquired for their retirement years. Following the destruction of his Indian garden "in the English taste" after a military defeat of the East India Company army, James Forbes carried the "seeds . . . preserved from the trees, shrubs, and flowers of Baroche, and different parts of Guzerat," back to England where many of them

have since flourished in the conservatory at Stanmore-hill. There I have the pleasure of beholding the tamarind-tree, custard-apple, and cotton-plant, flourishing with the ginger, turmeric, and coffee; and have gathered ripe guavas from a tree entwined by the crimson ipomea . . . encircled by the changeable rose . . . the fragrant mogree, attracting alhinna, and sacred tulsee.[76]

The "columns, urns, and other ornaments" in the *English* style that had embellished his garden in India before being "mutilated" by anti-imperial resistance, Forbes the retiree replaced by *Indian* architectural relics that "now adorn an octagon building at Stanmore-hill, erected for that purpose, under a linden-grove on the margin of a lake profusely adorned by the nymphea lotos, which . . . reminds me of the sacred tanks in Guzerat."[77] Plants, cultural artifacts, and aesthetic concepts from East and West were woven together into hybrid patterns, reflecting the ecocultural network that made such gardens possible in the first place.

Forbes was not the only returning nabob to inscribe a changed, dual, or hybrid cultural identity into his garden. As early as 1767, Captain John Gould had his Margate home designed "in imitation of a house in Calcutta."[78] After his return to the Scottish Highlands in 1782, Sir Hector Munro, ninth Commander-in-Chief of India, had a "Hindu temple" erected on his estate and his lawns further decorated with "faux ruins replicating the ancient gates of the city of Negapatam where he had won a celebrated battle during his career in India."[79] The first British mansion to be Indianized on its exterior was Warren Hastings's Daylesford House, designed by "architect Samuel Pepys Cockerell, brother of two nabobs," in 1788 with a Mughal dome.[80] A garden folly with "Indian" reminiscences was also erected by Reverend Jolland in Louth, Kent, in or before 1790, and consisted of an eccentric hermitage commemorating a brother who had died in India.[81] Another architectural object in the Indian style to decorate an English garden was a pavilion resembling a Hindu temple that Sir John Osborne had built at the turn of the century in Melchet Park, Wiltshire (now Hampshire), in honor of his friend Warren Hastings. Instead of an Indian god, however, the Melchet temple enshrined a Coade stone bust of the impeached former governor "emerging from a lotus flower" as a reincarnation of Vishnu. The temple was designed by Thomas Daniell, a British landscape painter who had traveled extensively in India; the bust was designed by John Rossi.[82]

Just like the Anglicized gardens in India, Indianized gardens in England could only grow out of the interactions of many players, including nonhuman organisms, individuals, and institutions across the globe, connected by networks of commerce, transport, communications, and knowledge. However, neither the hermitage in Kent, nor the Melchet pavilion nor Forbes's Stanmore temple survived the nineteenth century,[83] confirming Alan Lester's observation that "colonial networks . . . must be seen not only as provisional and contingent, but sometimes as ephemeral and even fleeting."[84]

The most durable and consistent manifestation of a fusion between English and Indian garden aesthetics survives in the greenscape surrounding Sezincote House, Gloucestershire, a country retreat remodeled in the "Indian manner" by Samuel Pepys Cockerell for his brother Charles, a former company man who had spent twenty-five years in and around Calcutta (Figure 3.1). Charles

FIGURE 3.1. South side of Sezincote House, Gloucestershire. Photograph by the author.

Cockerell had inherited the house from his older brother John, who died soon after returning to England from India. Remodeling began around 1805 and continued for more than fifteen years.[85] The artistic challenge at Sezincote consisted in creating a fitting backdrop for a residence that combined characteristics of Mughal and Hindu architecture[86] in a climate that was unsuitable for the growth of tropical plants out of doors. According to Desmond, the "Sezincote orangery was never particularly noted for its display of tender exotics" and very little information survives about the original outdoor plantings.[87] Kammerschmidt and Wilke mention several mighty cedars of Lebanon already in place before the Indian remodeling began,[88] some of which appear to have survived to the present day.[89] From visual clues in engravings by John Martin, Kammerschmidt and Wilke further deduce the earlier presence of islands of flowers and shrubs, inserted into the surrounding meadows and lawns. Conner claims that the "narrow valley" of the water garden "was planted with the most luxuriant vegetation available."[90] A recent newspaper article refers to an "ancient weeping hornbeam" that was "part of the original landscape planting."[91]

While the original plantings on the south side of the house followed formal patterns in the European tradition, the Indianization of the landscape on the north side took the form of a stone and water garden. Starting on a wooded hillside, water from a natural spring was directed into a number of pools, built at various levels and gradually descending down the slope to connect with a river at the bottom of the property (Figure 3.2).

Known as the "Thornery," the water garden was enhanced by a number of architectural components and topographical adjustments to evoke an Indian ambiance: A little temple shrine was dedicated to the Indian sun god Surya, in Vedic astrology associated with will-power, fame, general vitality, courage, kingship, and high social rank (Figure 3.3). Thanksgiving festivals in Surya temples established a close connection between the sun god and the food grain of wheat.[92] Surya's linkage to gold and brass as well as to the colors copper and red made him an excellent match for

FIGURE 3.2. Sezincote water garden; cascades. Photograph by the author.

Sezincote House with its warm yellow sandstone facades and copper dome. Below the shrine, a yoni-shaped "Temple Pool"[93] was created with a fountain column suggestive of the corresponding male symbol of fertility and regeneration, Shiva lingam.[94]

The pool was surrounded by a semicircle of artificial miniature grottos reminiscent of Indian cave temples. Several cascades further down, a "Hindoo" bridge on pillars, decorated with lotus

FIGURE 3.3. Sezincote water garden with Surya Temple, designed by Thomas Daniell. Photograph by the author.

motifs and sacred Brahmin bulls,[95] resembled bridge pillars near the Elephanta Caves of Bombay (Figure 3.4).

A stone-hewn meditation seat placed below the bridge could be reached by stepping stones placed in the water. On the east side of the bridge, a "Snake Pool" was decorated with a fountain in the shape of a three-headed metal snake hugging a tree trunk.[96] Still further below (and East), a "Rock Pool," surrounded by large boulders and dramatically stacked to suggest an Indian wilderness,[97] was framed with vegetation more typical of England.

The Sezincote Thornery is a prime example of the collaboration of individuals who belonged to overlapping private and professional networks related to the East India trade: Charles Cockerell, his deceased brother John, and his brother Samuel had all been official employees of the East India Company. Based on this professional connection, architect Samuel Pepys Cockerell had become a preferred choice for returning imperial officers, including Warren Hastings in nearby Daylesford,[98] who sought

FIGURE 3.4. Sezincote, Indian Bridge, designed by Thomas Daniell. Photograph by the author.

professional help with Indianizing traditional English country retreats on the outside, inside, or both.[99] For the garden design, the Cockerells additionally recruited landscape architect Humphry Repton, who had had his first encounter with an Indianized garden folly as early as 1790[100] and who would use his expertise gained at Sezincote to prepare designs in the Indian manner for the Prince Regent's Royal Pavilion at Brighton.[101] The exact degree of Repton's involvement in Sezincote remains uncertain, oscillating, according to different sources, between the submission of sketches for the overall design of the garden, selection of other artists' design elements, and contribution of his own architectural features to the garden.[102]

Because neither Samuel Pepys Cockerell nor Humphry Repton had ever visited India, both relied on the first-hand knowledge of a visual artist to augment the accuracy and authenticity of their work: Thomas Daniell, a painter who, together with his nephew William, had worked in India between 1784 and 1794 and, after their return, exhibited paintings with Indian motifs at the Royal Academy. The Daniells had also published *Oriental Scenery* (1795–1808), an aquatint collection that would influence British perceptions of India for centuries, with "naturalized" landscapes and "monuments of India" "distilled" into "a romantic vision of that country which has been absorbed to become a part of British life and culture."[103]

In Sezincote, the personal networks of company families, further intertwined by marital links,[104] architects, and visual artists converged: Samuel Pepys Cockerell's courage in integrating stylistic features of Indian architecture into English country house design was bolstered by his friendship with George Dance the Younger, "a rebellious and unconventional architect" who had set the example with London's Guildhall improvements of 1788. Dance had been influenced by Sir Joshua Reynolds's interest in Asian architecture, which in turn was inspired by the paintings that Hodges exhibited in London after his return from India. The "Gothick Architecture," Reynolds wrote in 1786,

> though not so ancient as the Grecian, is more so to our imagination with which the Artist is more concerned than with absolute truth. *The Barbarick splendor of those Asiatick Buildings*, which are now publishing by a member of this Academy [i.e., William Hodges], may possibly, in the same manner, furnish an

Architect, not with models to copy, but with hints of composition and general effect, which would not otherwise have occurred.[105]

Thomas and William Daniell had also visited India, encouraged by the successes of several other English painters including Johan Zoffany, Francis Swain Ward, and William Hodges. Hodges was especially well known to entertain a "high degree of connection" with other traveling painters.[106] This network of artists not only alleviated practical challenges related to life on the road in terrain still heavily contested between Eastern and Western political and military forces, it also contributed to the circulation of new painterly motifs and aesthetic ideas. Pictures of British expats posing in their opulent Anglo-Indian residences and gardens as well as of Indian monuments, cities, and landscapes sold well both in India and in Britain. Wealthy imperial administrators in India frequently became patrons and hosts of visiting English painters and invested part of their fortunes into art collecting long before retirement in Britain. Hodges, for example, stayed in India from 1781 to 1783 in response to a personal invitation by Governor Hastings. Many of Hodges's paintings from this period were later exhibited at the Royal Academy and, together with an aquatint series published as *Select Views in India*, "presented Indian architecture to British eyes for the first time as aesthetically admirable" and encouraged younger artists, including the Daniells, to follow his lead.[107] Charles Cockerell cultivated friendships with visual artists and commissioned seven additional oil paintings from Thomas Daniell as well as a series of engravings from John Martin (1789–1854) after the completion of Sezincote's Indian remodeling (Figure 3.5).[108]

Indianized mansions and garden follies triggered a temporary taste for Indian architecture in England;[109] they also stimulated the aesthetic debate of the day, carried out within a discursive network fed by art theorists, practicing artists, and architects. Their opinions on the value of ancient Indian architecture and its importance

FIGURE 3.5. *View of the Temple of Suryah & Fountain of Maha Dao, with a Distant View of North Side of Mansion House.* Drawn and etched by J[ohn]. Martin, engraved by F[rederick]. C[hristian]. Lewis, circa 1819. Courtesy, The Lilly Library, Indiana University, Bloomington, Indiana.

for the future of British art, however, remained irreconcilable. Sir Joshua Reynolds's guarded recommendation in his *Discourses*,[110] contrasted with James Malton's ridicule—

> There is no discrimination in the present style of Architecture. . . . The peculiars of every nation form a mongrel species in England: The rude ornaments of Indostan supersede those of Greece; and the returned Nabob, heated in his pursuit of wealth, imagines he imports the *chaleur* of the East with its riches[111]

—as much as with Humphry Repton's boundless enthusiasm:

> I cannot suppress my opinion that we are on the eve of some great future change in both these arts [i.e., "Gardening" and "Architecture"] in consequence of our having lately become acquainted with Scenery and Buildings in the interior provinces of India. The beautiful designs published by Daniell, Hodges

and other artists, have produced a new source of beauty, of elegance and grace, which may justly vie with the best specimens of Grecian and Gothic architecture.[112]

Sezincote remained a colorful exception instead of becoming a bridgehead of the new orientalized mainstream that Repton envisioned. Nevertheless, this Indianized house and garden survived into the twenty-first century and continue to celebrate their original owner's life-changing experience of transculturation. Both the orientalized garden and the orientalized garden owner acquired new identities owing to the movement of many goods and people through oriental networks (of commerce, knowledge, and art) that connected eighteenth-century Britain and India, and facilitated transformations in both places. The two-way movement of plants, botanists, imperial administrators, artists, and aesthetic concepts changed India as much as it changed Britain. Cockerell's garden would have remained an impossibility without cultural exchanges that facilitated new horticultural heterogeneity in both regions. Neither quite Indian nor entirely British, the water garden at Sezincote was the expression of a modulated taste and in turn it modulated the tastes of many visitors, including that of a prince who went on to transform the public face of a British spa town.[113]

Yet, despite its eye-catching exoticism, Sezincote also remained more closely connected to traditional English landscape gardens than it may seem at first sight. While Charles Cockerell was clearly eager to express his changed cultural identity, possibly also his admiration for Indian spirituality, the climatic difference between East and West limited his ambition as much as that of many other Anglo-Indian retirees. Unable to plant an Indian garden on English soil and obliged to rely mainly on architectural allusions, he created an "oriental" garden *ambiance* rather than a truly Indianized garden. In this respect he resembles Richard Boyle at Chiswick or Richard Temple at Stowe who, fifty years earlier, had recreated in their gardens the *ambiance* of classical antiquity (rather than an ancient Roman garden) to tell moral tales of ancient and modern political virtue.

Sezincote also resembles Stourhead, another garden adorned with monuments reminiscent of ancient architecture to tell a story, albeit a *literary* rather than a moral one: guided by visual and verbal clues, garden visitors were invited to reimagine the adventures of Aeneas and to share Henry Hoare's admiration for Virgil's *Aeneid*. All of these English landscape gardens express nostalgia for foreign cultures that their owners had encountered during their younger, most formative years. Both the garden designs of Stourhead and of Sezincote were inspired by personal travel memories and by highly idealized *visual* representations of the countries their owners had visited: Henry Hoare's three-dimensional fusion of ancient Rome and the Italy of his youth by the paintings of Salvator Rosa, Claude Lorrain,[114] and Gaspard Dughet;[115] Cockerell's romantic vision of India by the paintings of William Hodges (himself influenced by Lorrain),[116] Thomas and William Daniell.

In both places, Stourhead and Sezincote, garden temples dedicated to sun gods of ancient civilizations were placed in the most elevated topographical positions: Henry Hoare had his temple of Apollo erected on a hill top overlooking most other park buildings, cascades, and the lake; Charles Cockerell his temple of Surya overlooking a set of descending pools, themed fountains, and the Indian bridge along the Thornery slope. The analogous placement of Cockerell's Surya and the additional stylistic adaptation of Surya's relief and shrine to a European, neoclassical taste show the partial indebtedness of Sezincote garden also to a Western tradition of garden design and its awareness of and emulation with more traditional garden iconography in its vicinity (Figure 3.6).

Hosting an oriental rather than an occidental deity in his garden, yet having the Hindu god's appearance adjusted to an aesthetic modeled on representations of ancient Greek gods (through the removal of Surya's third eye and second set of arms),[117] Cockerell inscribed into his garden ornament his fascination with Eastern culture and his simultaneous adherence to his Western cultural heritage. He also created a transculturated space where seemingly familiar objects obtain new cultural connotations that they previously possessed neither in India nor in Britain. As an

FIGURE 3.6. Sezincote water garden; detail of Surya Temple. Photograph by the author.

innovative, self-consciously deviating response to Stourhead's Temple of Apollo, possibly also to Chiswick's Ionic Temple and Stowe's Temple of Venus, Sezincote's Temple of Surya, then, reveals yet another, more subtle communication network among the gardens themselves, allowing East and West to speak, through garden art, directly to one another.

Notes

1. See "network, n.," *Oxford English Dictionary Online*, 2nd ed. (Oxford: Oxford University Press, 1989), http://www.oed.com/.

2. James Beattie, Edward Melillo, and Emily O'Gorman, "Introduction," in *Eco-Cultural Networks and the British Empire: New View on Environmental History*, ed. James Beattie, Edward Melillo, and Emily O'Gorman (London: Bloomsbury, 2015), 8.

3. Beattie, Melillo, and O'Gorman, *Eco-Cultural Networks*, 7. The authors examine environmental interactions within and between empires and describe "eco-cultural networks" as networks consisting of individuals from diverse cultures, institutions, and nonhuman organisms around the world that are drawn together by imperial interests. Empire building facilitates mutual exchanges of knowledge, cultural practices and materials (including plants and animals) with intentional or unintentional effects on the cultures, markets, and ecosystems within imperial influence zones as well as beyond their territories.

4. See Penelope Hobhouse, *Gardening through the Ages: An Illustrated History of Plants and Their Influence on Garden Styles from Ancient Egypt to the Present Day* (New York: Barnes and Noble, 1997), especially chapters 3–5.

5. Ray Desmond, *The European Discovery of the Indian Flora* (Oxford: Oxford University Press, 1992), especially 33–50.

6. Quoted from Michael H. Fisher, ed., *Visions of Mughal India: An Anthology of European Travel Writing* (London: I. B. Tauris, 2007), 152–153.

7. Fisher, *Visions of Mughal India*, 189 and 195–196.

8. Michael H. Fisher, ed., *The Travels of Dean Mahomet: An Eighteenth-Century Journey through India* (1794) (Berkeley and Los Angeles: University of California Press, 1997), 34.

9. Fisher, *Travels of Dean Mahomet*, 86.

10. For the garden trope as a commonality between India and England, see Lynn Staley, *The Island Garden: England's Language of Nation from Gildas to Marvell* (Notre Dame, IN: University of Notre Dame Press, 2012). For England as a garden, see also Joseph Addison, *Tatler*, 218 (August 31, 1710), in Joseph Addison and Richard Steele, *The Tatler*, ed.

Donald F. Bond (Oxford: Clarendon, 1987), 3:159. Two hundred years later, Rudyard Kipling would expand the idea in his poem "The Glory of the Garden" (1911).

11. John Milton, *Paradise Lost*, ed. Scott Elledge (New York: W. W. Norton, 1975), 210 (book 9, ll. 1101–1103).

12. "An Epistle from a Lady in England to a Gentleman at Avignon," *The Poetical Works of Thomas Tickell* (Edinburgh: Apollo, 1781), 116, ll. 185–190.

13. Robert Southey, *The Curse of Kehama* (London: Longman, 1810), part 13, 133–134.

14. Private collection; reprinted in Hermione De Almeida and George H. Gilpin, *Indian Renaissance: British Romantic Art and the Prospect of India* (London: Ashgate, 2005), 12.

15. Yale Center for British Art, Paul Mellon Collection.

16. Reprinted in *British Encounters with India, 1750–1830: A Sourcebook*, ed. Tim Keirn and Norbert Schürer (Basingstoke: Palgrave Macmillan, 2011), 211.

17. Thomas Daniell and William Daniell, *Oriental Scenery: 150 Views of Architecture, Antiquities and Landscape Scenery of Hindostan* (London, 1795–1816).

18. James Forbes, *Oriental Memoirs: Selected and Abridged from a Series of Familiar Letters Written During Seventeen Years Residence in India: Including Observations on Parts of Africa and South America, and a Narrative of Occurrences in Four India Voyages; Illustrated by Engravings from Original Drawings*, 4 vols. (London: White, Cochrane, and Co., 1813).

19. Bärbel Czennia, "Nationale und kulturelle Identitätsbildung in Grossbritannien, 1660–1750. Eine historische Verlaufsbeschreibung," in *Muster und Funktionen kultureller Selbst- und Fremdwahrnehmung. Beiträge zur internationalen Geschichte der sprachlichen und literarischen Emanzipation*, ed. Ulrike-Christine Sander and Fritz Paul (Göttingen: Wallstein, 2000), 355–390.

20. *The Seasons / Summer*, in *The Complete Poetical Works of James Thomson*, ed. J. Logie Robertson (Oxford: Oxford University Press, 1908), 84, ll. 861–887.

21. Part V, Thomson, *Complete Poetical Works*, 392–393, ll. 9–32.

22. *The Poems of Alexander Pope: A One-Volume Edition* [1963], ed. John Butt (New Haven, CT: Yale University Press, 1977), 195–210, ll. 29–32 and

365. For similar rhetoric see also William Mason on a "wise Sidonian"'s garden in *The English Garden: A Poem. In Four Books* (Dublin: S. Price et al., 1782), Book the Second, 59–63.

23. Thomson, *Summer, Complete Poetical Works*, 78, ll. 663–681.

24. Pope, *Windsor-Forest*, ll. 29–32.

25. According to the website of the Twickenham Museum, the tree had been available from English nurseries since the 1730s ("Alexander Pope's Willow Tree: A Weeping Legend," http://www.twickenham -museum.org.uk/detail.php?aid=287&ctid=2&cid=1.

26. *The Newcastle General Magazine, or Monthly Intelligencer* 1 (1748): 25–28; later reprinted in Maynard Mack, *The Garden and the City: Retirement and Politics in the Later Poetry of Pope, 1731–1743* (Toronto: Toronto University Press, 1969), 237–243.

27. *Poems*, 619–624, ll. 137–148.

28. Paul Baines, "Alexander Pope," in *The Cambridge Companion to English Poets*, ed. Claude Rawson (Cambridge: Cambridge University Press, 2011), 249.

29. Andrea Wulf, *The Brother Gardeners: Botany, Empire, and the Birth of an Obsession* (New York: Knopf, 2009), 256, states that the North American or black walnut (*Juglans nigra*) had been cultivated in Britain since the seventeenth century but still counted as a rarity in 1731. Wulf also lists several laurels introduced to Britain from America during the eighteenth century.

30. [Joseph Addison], *Spectator* 69 (May 19, 1711), in Joseph Addison and Richard Steele, *The Spectator*, ed. Donald F. Bond (Oxford: Clarendon Press, 1965), 1:295–296.

31. For changing connotations of the technical terms "hothouse," "conservatory," "greenhouse," and "stove[-house]," in eighteenth-century English usage, see Desmond, *European Discovery*, 294–300. Both Desmond (291) and Andrea Wulf and Emma Gieben-Gamal, *This Other Eden: Seven Great Gardens and 300 Years of English History* (London: Little Brown, 2005), 64, quote an eyewitness who had admired 400 rare Indian plants on display in the royal stove-houses at Hampton Court as early as 1690.

32. Wulf, *Brother Gardeners*, 4.

33. Philip Miller, *The Gardeners Dictionary: Containing the Methods of Cultivating and Improving the Kitchen, Fruit and Flower Garden; as also, the Physick Garden, Wilderness, Conservatory, and Vineyard; According to the Practice of the Most Experinc'd Gardeners of the Present Age* (London, 1731), unpaginated edition, especially letters "CO," "JA," "LI," and "ST."

34. William Aiton, *Hortus Kewensis; Or, A Catalogue of the Plants Cultivated in the Royal Botanic Garden at Kew. By William Aiton, Gardener to His Majesty* (London, 1789), vol. 1: 41, 93, 96, 106, 115, 154, 156, 260, 286, and 318; vol. 2: 36, 54, 59, 77, 133, 134, 137, 321, 360, 438, 442, 444, 453, and 455; vol. 3: 29, 31, 34, 63, 69, 71, 81, 83, 101, and 102. In addition to the geographical origin of each plant, Aiton provides the names of people who first introduced or successfully cultivated the species in England. Aiton's lists repeatedly mention the botanists Sir Joseph Banks and Daniel Charles Solander, professional gardeners Philip Miller and John Walsh, as well as the wealthy hobby gardener "Dutchess [*sic*] of Beaufort."

35. Aiton, *Hortus Kewensis*, 1: 45, 115, 154, and 260.

36. Aiton, *Hortus Kewensis*, 2: 321 and 453.

37. Aiton, *Hortus Kewensis*, 3: 29, 69, and 102.

38. John Abercrombie, *The Propagation and Botanical Arrangements of Plants and Trees, Useful and Ornamental, Proper for Cultivation in Every Department of Gardening* (London 1784), 2: 745, 752, 789, 797, and 807.

39. Elisabeth Lenckos, "Daylesford," *The East India Company at Home, 1757–1857*, August 2014, http://blogs.ucl.ac.uk/eicah/files/2014/08/Daylesford-PDF-Final-20.08.141.pdf.

40. Desmond, *European Discovery*, 294–97. Banks's use of personal networks in his pursuit of Indian plants is also documented in various inventories of "Boxes Chests & Baskets of plants sent out from the Botanic Garden" [of Calcutta] as well as in private letters reprinted in *The Indian and Pacific Correspondence of Sir Joseph Banks, 1768–1720*, ed. Neil A. Chambers, vol. 8: *Letters 1810–1821* (London: Pickering and Chatto, 2008–2014), 192–193 and 217–219.

41. Quoted in Desmond, *European Discovery*, 273, and also discussed in Eugenia W. Herbert, *Flora's Empire: British Gardens in India* (Philadelphia: University of Pennsylvania Press, 2011), 69.

42. For the practical challenges of keeping tropical plants alive and seeds viable aboard saltwater-drenched sailing ships, which took between four and six months for a one-way trip between England and India, see Desmond, *European Discovery*, chap. 21 ("Transportation of Plants") and Wulf and Gieben-Gamal, *This Other Eden*, 232–233.

43. The term "marigold" has been used for flowers in the *Tagetes* family (originating in Aztec South America) but also for *Calendula officinalis*, a southern European species. Hobhouse, *Gardening*, 96–135, emphasizes the special importance of the Renaissance period for the introduction to England of many new species that would later be regarded as "English."

44. While homesick British expatriates doubtless associated these flowers with their beloved homeland, many plants perceived as "English" by eighteenth-century gardeners had only recently been naturalized in the British Isles: Sweet William (*Dianthus barbatus*), for example, comes in two varieties one of which originated in southern Europe (*Dianthus barbatus var. barbatus*), the other in Asia (*Dianthus barbatus var. asiaticus Nakai*). According to J. H. Krelage, "On Polyanthus Narcissi," *Journal of the Royal Horticultural Society* 12 (1890): 339–346, polyanthus originated in the Mediterranean region and was mainly cultivated in Holland before reaching England.

45. Herbert, *Flora's Empire*, 65. In the interest of consistency, eighteenth-century place names will be used for Indian locations throughout this essay: for example, "Calcutta" for modern-day Kolkata and "Bombay" for modern-day Mumbai.

46. Eustace Carey, *Memoir of William Carey, D.D.* (London: Jackson and Walford, 1836), 507.

47. For botanic gardens as nodes within eco-cultural networks, see also Beattie, Melillo, and O'Gorman, *Eco-Cultural Networks*, 9, and Eugenia W. Herbert, "Peradeniya and the Plantation Raj in Nineteenth-Century Ceylon," in the same volume, 123–150.

48. Wulf, *Brother Gardeners*, 207–221.

49. Charles Carlton and Caroline Carlton, "Gardens of the Raj," *History Today* 46, no. 7 (1996), 26.

50. For the genesis of the Calcutta garden as well as its strong historic ties to the botanical garden of Cape Town, see Desmond, *European Discovery*, 57–63.

51. Wulf, *Brother Gardeners*, 221.

52. Wulf, *Brother Gardeners*, 209ff., and Desmond, *European Discovery*, 61ff.

53. Wulf, *Brother Gardeners*, 211, 215, and 217.

54. Niklas Thode Jensen, "The Tranquebarian Society: Science, Enlightenment and Useful Knowledge in the Danish-Norwegian East Indies, 1768–1813," *Scandinavian Journal of History* 40, no. 4 (2015), 535–561, offers a detailed comparative analysis of research activities in Tranquebar, Calcutta, and Batavia; he also describes their collaborative efforts as an early modern form of regional networking.

55. Jensen, "Tranquebarian Society," 535, 536, 539, and 541.

56. Desmond, *European Discovery*, 39 and 52–53. As early as 1700, the Royal Society in London had received nearly 400 plants from India through the East India Company, described by Sam. Brown and James Petiver in several instalments of *Philosophical Transactions of the Royal Society of London* 22 (1700–1701): 579–594; 699–721; 843–862; 927–946; 1007–1029; and 23 (1702–1703): 1055–1068, 1251–1266 [last page incorrectly numbered as 1566], 1450–1460.

57. Desmond, *European Discovery*, 298ff.

58. Benedict Anderson, *Imagined Communities: Reflections on the Origin and Spread of Nationalism* (London: Verso, 1991), 6–7. For the outstanding service of Roxburgh and his successor, Wallich "to the needs and requests of British gardeners," see also Desmond, *European Discovery*, 300.

59. Collection of the Victoria Memorial Hall, Calcutta, India.

60. Date unknown; Maidstone Museum and Bentlif Art Gallery, Kent; for additional comment, see Carlton and Carlton, "Gardens," 23.

61. *Travels*, 59 and 97–98.

62. See also C. and C. Carlton, "Gardens of the Raj," 23: "Jungle was cleared and marshes were drained to create a Georgian landscape. Using techniques developed by . . . Capability Brown, a vista was created of the church steeple at Serampan," a former Danish settlement across the river.

63. Herbert, *Flora's Empire*, 69–70.

64. Raymond Head, *The Indian Style* (London: Allen and Unwin, 1986), 22; Carlton and Carlton, "Gardens of the Raj," 23; and Giles H. R. Tillotson, *The Artificial Empire: The Indian Landscape of William Hodges* (Richmond: Curzon, 2000), 70ff.

65. Carlton and Carlton, "Gardens of the Raj," 23, quote from a letter of Lady Canning to Queen Victoria: "The Park is so carefully planted with round headed trees to look as English as possible."

66. For a persistently "high failure rate" of imported European trees, especially "European fruit trees," see Desmond, *European Discovery*, 59.

67. National Portrait Gallery, London.

68. De Almeida and Gilpin, *Indian Renaissance*, 135.

69. Forbes, *Oriental Memoirs*, 2:239–240, my italics.

70. Forbes, *Oriental Memoirs*, 3:138.

71. Forbes, *Oriental Memoirs*, 2:240–241.

72. Forbes, *Oriental Memoirs*, 2:241–242.

73. Forbes, *Oriental Memoirs*, 3:408. "Artlessness" in this context echoes Joseph Addison's redefinition of the English garden in contradistinction to the formal baroque garden, for instance, in *Tatler* 161 (April 20, 1710) and 218 (August 31, 1710), in Addison and Steele, *The Tatler*, ed. Bond, 2:397–401 and 3:140–143. See also *Spectator* 414 (June 25, 1712), in Addison and Steele, *The Spectator*, ed. Bond, 3:548–553. See also Alexander Pope's raillery against topiary in *Guardian* 173 (September 29, 1713), in Joseph Addison and Richard Steele, *The Guardian*, ed. John Calhoun Stephens (Lexington: University of Kentucky Press, 1982), 562–565.

74. "Lines inscribed under an Urn in a Garden at Baroche, near a Spring overshadowed by a Burr, or Banian-tree, surrounded by flowering Shrubs," *Oriental Memoirs*, 2:242–243.

75. *Nabobs: Empire and Identity in Eighteenth-Century Britain* (Cambridge: Cambridge University Press, 2010), 238. In addition to Indian plants, garden concepts, and mythological allusions, James Forbes also adopted elements of an "Indian" lifestyle, including the local bathing culture, to which he attributed his resilience vis-a-vis tropical diseases. *Oriental Memoirs*, vol. 2, 410–411.

76. Forbes, *Oriental Memoirs*, 3:409.

77. Forbes, *Oriental Memoirs*, 3:361–362. According to Patrick Conner, *Oriental Architecture in the West* (London: Thames and Hudson, 1979), 119, the Stanmore temple was built sometime between 1784 and 1793, probably making it the earliest English garden ornaments in the Indian manner.

78. Tillman W. Nechtman, *Nabobs: Empire and Identity in Eighteenth-Century Britain* (Cambridge: Cambridge University Press, 2010), 167.

79. Nechtman, *Nabobs*, 167.

80. Head, *Indian Style*, 12–13.

81. Valentin Kammerschmidt and Joachim Wilke, *Die Entdeckung der Landschaft: Englische Gärten des 18. Jahrhunderts* (Stuttgart: Deutsche Verlagsanstalt, 1990), 161–162.

82. Conner, *Oriental Architecture*, 117–119; Head, *Indian Style*, 13–14; Desmond, *European Discovery*, 296; Mildred Archer, *Early Views of India: The Picturesque Journey of Thomas and William Daniell, 1786–1794* (London: Thames and Hudson, 1980), 230. An early description of the garden monument was published anonymously as "Some Account of a Hindu Temple, and a Bust, of which Elegant Engravings Are Placed in the Oriental Library of the Hon. East India Company, in Leadenhall Street. [With Two Plates.]," *European Magazine* 42 (December 1802): 448–449.

83. Conner, *Oriental Architecture*, 119.

84. Alan Lester, "Imperial Circuits and Networks: Geographies of the British Empire," *History Compass* 4, no. 1 (2006), 135.

85. Because the focus of this essay lies on gardening, I refer readers interested in details of the architectural remodeling of Sezincote House to Conner, *Oriental Architecture*, 120–126; Head, *Indian Style*, 14, 16, 37–49; Archer, *Early Views*, 230–233; Kammerschmidt and Wilke, *Entdeckung*, 162–170; and Jan Sibthorpe's case study "Sezincote, Gloucestershire" for *The East India Company at Home, 1757–1857*, May 2014, https://cpb-eu-w2.wpmucdn.com/blogs.ucl.ac.uk/dist/1/251/files/2013/06/Sezincote-PDF-Final-19.08.14.pdf.

86. Most sources emphasize the onion-shaped dome, four small *chatris* or minarets, and deep overhanging eaves, called *chujjahs*, as architectural features in the Mughal tradition, and pillars, ubiquitous lotus representation, and "peacock tail" window arches as Hindu heritage; many features are echoed in adjacent structures, including smaller lodges and farm buildings on the same property.

87. Desmond, *European Discovery*, 297.

88. Kammerschmidt and Wilke, *Entdeckung*, 170.

89. I thank the current owners of Sezincote House and Gardens, Edward Peake and his family, for allowing me to visit and take photographs outside regular opening hours in June 2016.

90. Conner, *Oriental Architecture*, 122.

91. Val Bourne, "Sezincote Garden, Gloucestershire: A Passage to India," *The Telegraph*, July 6, 2012, https://www.telegraph.co.uk/gardening /9379639/Sezincote-garden-Gloucestershire-a-passage-to-India.html.

92. Heinrich Zimmer, *Myths and Symbols in Indian Art and Civilization*, ed. Joseph Campbell (Princeton, NJ: Princeton University Press, 1972).

93. By June 2016, Sibthorpe's epithet "yoni-shaped" ("Sezincote," 18) appeared more accurate than Kammerschmidt and Wilke's comparison of the pool shape to the contours of a lotus flower (*Entdeckung*, 168). Head's adjective "circular" (*Indian Style*, 40) corresponds to an oil painting by Thomas Daniell, *Temple, Fountain and Cave in Sezincote Park* (1819; Yale Center for British Art, New Haven, CT), suggesting that the pool shape may have changed over time.

94. For the symbolic connotations of lingam and yoni in relation to Shiva, Lord of Lingam, Vishnu, supreme creator of the universe, his wife Lotus, and other Hindu deities, see Zimmer, *Myths and Symbols*, 123–130. Kammerschmidt and Wilke, *Entdeckung*, 168, point out further symbolic references to the idea of rebirth. Desmond, *European Discovery*, 25, traces Surya back to the ancient Indian vedas, an important source of "[c]urrent knowledge of botany in ancient India."

95. Sculptures of the bull Nandi, perceived either as Shiva's vehicle of transport or his animal representation, were often positioned as gatekeepers near the god's temples and shrines.

96. For multiheaded serpents, serpents more generally as subaquatic keepers of life-energy, and serpents as representation of Vishnu, see Zimmer, *Myths and Symbols*, 63, 75, and 78ff.

97. Owing to vegetation growth, the boulder stacks are no longer visible today. For additional water features on the south side of the house, including an eighteenth-century grotto with an Indian water maze and a modern recreation of a Mughal-style *charbagh*, see Sibthorpe, "Sezincote," 28.

98. According to Kammerschmidt and Wilke, *Entdeckung*, 162, and Lenckos, "Daylesford," 1–2, orientalizing features of Hastings's

Daylesford were limited to the house itself: a Mughal-style central dome on the exterior, fireplace frescoes, furniture, and art objects on the interior. For surviving nonorientalized features, including several "mature trees," the "orangery," and smaller architectural components by John Davenport, see "Daylesford House," Historic England, https://www.historicengland.org.uk/listing/the-list/list-entry/1000760.

99. For more details on returning nabobs "remaking their homes into micro-Indias," see Nechtman, *Nabobs*, 166–170.

100. Kammerschmidt and Wilke, *Entdeckung*, 162.

101. An Indianized pavilion was to complement the Prince of Wales's "Indian" stables, designed by William Porden, a pupil of Samuel Pepys Cockerell, in 1802. Repton's designs were later rejected in favor of those by John Nash.

102. In the "Prefatory Observations" to his *Designs for the Pavillon at Brighton* (London: J. C. Stadler, 1808), v, note f, Repton himself concedes that he gave his "opinion concerning the adoption of this new style, and even assisted in the selecting some of the forms from Mr. T. Daniell's collection"; but neither does Repton claim credit for "the architectural department" nor for any landscaping at "Sesincot"; ditto Stephen Daniels, *Humphry Repton: Landscape Gardening and the Geography of Georgian England* (New Haven, CT: Yale University Press, 1999), 194, in turn referring to Christopher Hussey, *English Country Houses: Late Georgian 1800–1840* (London: Country Life, 1958), 66–73. By contrast, Nechtman, *Nabobs*, 168, claims that Repton designed "a series of garden follies that included a Hindu temple, a bridge guarded by two sacred cows, and a mock-ruin designed to replicate a snake pool." This contradicts Archer, *Early Views*, 231, and Head, *Indian Style*, 38–39, who both argue that Daniell designed the garden monuments, referencing detailed sketches and letters of Daniell to Charles Cockerell. Both Conner's allusion to a "preliminary sketch in grey wash" in *Oriental Architecture*, 121, and Head's allusion to a "surviving drawing" in *Indian Style*, 38, suggest that Repton perhaps also influenced or at least intended to influence the overall garden design. Head, *Indian Style*, 39, further credits Repton for "the idea for a conservatory," for "altering the farmhouse on the hill opposite the south front to a *serai*," for "the positioning of the octagonal

fountain," for "the gardener's house, which was to be constructed like a Hindu temple among some trees, and the siting of a grotto."

103. According to Archer, *Early Views*, 233, "[e]ven today . . . the popular vision of India still remains that created by *Oriental Scenery*."

104. Sibthorpe, "Sezincote," 18, refers to multiple intermarriages between the Cockerell and Hastings families. Professional loyalties were frequently intensified by private connections within the nabob community.

105. My italics. Joshua Reynolds, "Discourse XIII" (1786), in *Discourses on Art*, ed. Robert R. Wark (New Haven, CT: Yale University Press, 1975), 242. For the effect of Reynolds's ideas on Dance's 1788 revision of his plans for the London Guildhall, see Head, *Indian Style*, 23–24.

106. Tillotson, *Artificial Empire*, 69.

107. Conner, *Oriental Architecture*, 114. He also quotes Sir William Jones, *Asiatick Researches* I (1788), 411, as predicting that "correct delineations" of Indian buildings would soon "furnish our own architects with new ideas of beauty and sublimity," thereby echoing Sir Joshua Reynolds's "Discourse XIII" of 1786 (see note 105).

108. Archer, *Early Views*, 231, and Head, *Indian Style*, 43. John Martin, however, had not visited India.

109. Conner, *Oriental Architecture* as quoted above; Mildred Archer, "The Daniells in India and their Influence on British Architecture," *RIBA Journal* 67 (1960), 444; Mildred Archer, *Early Views*, 233; more generally also Head, *Indian Style*, as quoted above (note 108).

110. See Reynolds as quoted earlier (note 105).

111. James Malton, *An Essay on British Cottage Architecture: Being an Attempt to Perpetuate on Principle, that Peculiar Mode of Building, which was Originally the Effect of Chance . . .* (London: Hookham and Carpenter, 1798), 9–10. Head, *Indian Style*, 23, additionally quotes Thomas Sandby, a professor of architecture at the Academy who regarded "the wild and extravagant" forms of Indian architecture as having "nothing but curiosity value" and "no relevance for English architects."

112. Humphry Repton, *An Enquiry into the Changes of Taste in Landscape Gardening* (London: J. Taylor, 1806), 41. According to Daniels, *Humphry Repton*, 197, Repton not only regarded ancient Indian architecture "as an advance on Grecian and Gothic styles" and as

harbinger of stylistic change but "went further to sketch a theory proposing that forms of Indian architecture correspond to forms of flowers, as Gothic forms corresponded to buds and Grecian to leaves."

113. Conner, *Oriental Architecture*, 131–153; Head, *Indian Style*, 38–39 and 44–46.

114. Hans von Trotta, *Der Englische Garten: Eine Reise durch seine Geschichte* (Berlin: Wagenbach, 1999), 56–57.

115. Linda Cabe Halpern, "The Uses of Paintings in Garden History," *Garden History, Issues, Approaches, Methods*, ed. John Dixon Hunt (Washington, DC: Dumbarton Oaks Research Library and Collection, 1992), 190.

116. Head, *Indian Style*, 22.

117. For the normalization of Surya at Sezincote and parallels in the placement of Apollo's and Surya's temples in Stourhead and Sezincote, see also Kammerschmidt and Wilke, *Entdeckung*, 168.

4

The Blood of Noble Martyrs

Penelope Aubin's Global Economy of Virtue
as Critique of Imperial Networks

SAMARA ANNE CAHILL

In the eighteenth century the concept of the "network" was evolving from an older, sixteenth-century understanding of "network" as a material craftwork, something "in the fashion of a net" created by man or God, to the nineteenth-century understanding of "network" as a "chain or system of interconnected immaterial things."[1] One such "chain or system" was Penelope Aubin's imaginary network of reformed and reunified Christianity, disseminated across the world through human "nodes" of aristocratic martyr figures. Aubin (1679?–1738)—a novelist who has recently experienced a surge in scholarly interest—was critical of England's imperial ambitions, seeing them, I argue, as evidence of its reliance on foreign luxury. Aubin's critique of luxury focused particularly on the Muslim-dominated section of the trade network of the Silk Road that extended from the Mediterranean to Tartary, but she indirectly also criticized Britain's own ambitions as an emerging new empire of commerce. By contrasting noble Christian martyrs with a stereotypically imagined Islam, Aubin portrayed her altruistic network as a truly global, alternate economy that could replace the network of corrupt imperial trade.

In Aubin's view, her fellow Christians in England and abroad had failed in their duty to support each other against the forces of atheism and libertinism. She berated the "abominable writings" of John Toland, accused *Robinson Crusoe* of being less probable than her own narratives, and wished that the

> English would be again (as they were heretofore) remarkable for Virtue and Bravery, and our Nobility make themselves distinguished from the Crowd, by shining Qualities, for which their Ancestors became so honoured, and for Reward of which obtained those Titles they inherit.[2]

Aubin saw the moral failure of England's reading public reflected in the popularity of Toland's writing and Defoe's fiction and in the failure of the nobility to imitate the sacrificial example of their more idealistic ancestors. In turn, she portrayed the luxurious trade network in the Ottoman Mediterranean as a dangerous space that promised worldly pleasure only to threaten the spiritual bonds among Christians; the sacrifice and endurance of members of the nobility alone could recuperate Christian unity. Flung apart by the ignoble drives of luxury, material pleasure, and trade, Christians suffer loss, hardship, injury, and dislocation before being brought back together by sacrifice that is rewarded with providential deliverance.

Dwight Codr has recently argued that English fiction in the period 1690–1750 mediated the opposed rhetoric of providentialism (trust in God's omniscience and omnipotence) and of financial risk (how to minimize one's financial losses in an uncertain world) through various responses to "usury," which he sees as the attempt to play God in finances by achieving maximum profit with minimum risk.[3] My argument relies on Aubin's investment in the "providentialist" coin: the continuing tradition of rejecting material wealth for that "higher economy in which providence . . . was paramount."[4] In this, too, Aubin saw herself as opposed to Daniel Defoe, her competitor in the realm of fiction. Of their travel fiction, Bannet points out that "Aubin repeatedly contested" Defoe's mercantile-derived "representations of the Atlantic."[5] I would argue that she also rejected his representations of risk and profit. If Defoe defended entrepreneurial projects—Robinson Crusoe is, in many ways, a consummate projector as well as a Christian British imperialist—in "a climate in which anticipatory planning or scheming was relentlessly associated with ethical, epistemological, and spiritual deviance,"[6] then Aubin reaffirmed the deviance of not trusting to God unto death.[7] A noble martyr's voluntary shedding of blood is the ultimate act of putting one's money (in the sense of a moral countercurrency) where one's mouth is, of risking everything in the world on the promissory note of Providence.

1

Aubin's critical reputation has experienced a minor renaissance of late. Due to the fine scholarship of Debbie Welham, the scholarly community now has a portrait of Aubin as an Anglican (probably a High Anglican with Tory and Royalist allegiances), an orator, a woman with ties to trade and (like Jonathan Swift) to the Temple family, and a robust interest in the complexities of British national identity. Welham's investigation of the wills and correspondences of the Aubin, Temple, and Charleton families enabled her to conclude that Penelope Aubin was "the half-sister of Viscount Cobham, illustrious owner of Stowe, friend of Alexander Pope and a patron of the Whig party."[8] Recent scholarship by Eve Tavor Bannet, Edward Kozaczka, Chris Mounsey, and Adam Beach has further solidified Aubin's burgeoning reputation as a novelist deeply interested in situating England on a global stage.[9] Given the complexity of Aubin's global perspective as it was informed by a steely and unwavering conviction in Christian superiority, the concept of the network is useful in assessing Aubin's attempt to reconcile the distant with the domestic in response to England having two kings on foreign land simultaneously in the 1720s: James Francis Edward Stuart, considered by Jacobites to be the true king of England, lived in France; George I, who succeeded to the throne in 1714 due to the exclusion of Roman Catholics by the Act of Settlement, was not born in England and preferred to return to his native Hanover whenever possible. This shared sense of monarchical displacement among Britons, whether they affiliated themselves with the Jacobite or Hanoverian king, meant that representing the distant and the foreign could also be a negotiation of what constituted the domestic. The network—a metaphorical map of relationships in which the distant and the domestic are imbricated and mutually informing—is thus an apt heuristic for reading Aubin's fiction.

Aubin depicted British identity as a hybrid Christianity threatened by the corrupting network of global trade. Welham's evidence that Aubin held an Anglican allegiance alongside a cosmopolitan awareness of the rewards and risks of trade would explain why Aubin envisioned British identity in such a unique way. Indeed, Aubin depicted the complexities of pro-Revolution and pro-Jacobite British identity as reconcilable, but only by construing the material pleasures of an imaginary (and stereotypical) Muslim East as an unassimilable temptation to Christian unity. Though Aubin did not use the term "network," she constructed Christian unity—whether European or specifically British—as a moral network held together at the nodes of martyrial example. From *The Strange Adventures of the Count de Vinevil and His Family* (1721), one of her earliest novels, through *The Life and Adventures of the Young Count Albertus* (1728), her final novel, Aubin used the hybrid identities of noble men to juxtapose the temptations of "Eastern" trade with the moral unity of European Christians who unite around an aristocratic martyr's example.

The two eponymous heroes nevertheless show a development in Aubin's work across the 1720s. Count Vinevil, though a French nobleman, seems a stand-in for exiled Jacobites. Vinevil is an aristocrat who resents being displaced by favorites in the government, abandons his home country, and rashly turns to trading with the Ottoman Empire to achieve the success in trade that the court denied him in preferment. As a voluntary exile he is depicted as a sympathetic character who understandably, but wrongly, rejected his homeland for worldly success. Realizing his error too late, he nobly sacrifices himself while defending the virtue of his daughter Ardelisa from the sexual predation of Mahomet, the Bassa of Constantinople's son. Through Vinevil's example, Ardelisa and her husband, Count Longueville, survive to gather around themselves an increasingly diverse band of European aristocrats. Their affective Christian bonds are formed in opposition to the "Empire of the Eastern World."[10]

While Count Vinevil is a stand-in Jacobite, Aubin's final hero, Count Albertus, is the hybrid child of a Williamite and a Jacobite: he literally embodies British Christian unity. Albertus is also, through his global travels, the moral center of a band of Christian castaways and exiles in the Mediterranean. Ultimately, he journeys to China as a missionary where he is captured, taken to Tartary, and eventually martyred. Aubin's novels have been criticized for their multilayered, repetitive plots and their episodic nature. But Aubin employs the ensuing iterations—particularly the complicated embedded narratives of Christian exiles and castaways in Ottoman and Barbary locations such as Constantinople and Tunis—to depict a partly imagined Muslim East as so entirely a threat to Christianity that it brings about an otherwise unachievable Christian unity. One example is the self-sacrificial death of Count Albertus, a human node within the moral network that Aubin promoted in her prefaces and fiction as pious alternative to the lucrative trade network. Further, in *The Life of Charlotta du Pont, an English Lady* (1723) Aubin's narrator reinforces a narrative of Christian unity in the face of worldly (and often stereotypically Muslim) predation by stepping out of the narrative to comment,

> 'tho we often live as ill as Heathens, who profess our selves
> Christians, and whilst we live together are often at variance; yet
> none but such as have experienced it, can tell the Joy and
> Comfort poor Christians find, in meeting and conversing
> together when in Slavery, and amongst Turks and Heathens.[11]

The pleasures of global trade tempt disunited Christians, but the violence of "Eastern" empires toward noble martyrs brings these

Christians back together in a moral network that extends over the whole world. In a period shortly before the British Empire actually became a global superpower, Aubin represents British Christian unity as achieving a moral superiority that rises in the East and extends to the West, traversing the whole world.

2

Penelope Aubin was deeply concerned about the relationship between fiction, national morality, and national integrity. She saw Christian identity as a bulwark of national integrity and her novels were thus moral and political projects of collective education. In a prefatory letter included in volume 3 of the 1739 edition of her collected works, she explicitly distanced herself from what she considered to be the immoral and frivolous crowd of "other female Authors."[12] Further, the appeal of Aubin's novels, according to the anonymous writer of another preface (possibly Samuel Richardson), originated in her pious conclusion that "Religion is no Jest, Death and a future State certain." The preface suggests that her novels will "inspire in us" the "noble Sentiments" which are absent from "loose Writings which debauch the Mind."[13] For Aubin, virtuous self-denial was an index of national flourishing.

Aubin herself explains that she wrote to "encourage Virtue, and excite us to heroic Actions" at a time when the court "being removed to the other side of the Water, and beyond Sea, to take the Pleasures this Town and our dull Island cannot afford; the greater part of our Nobility and members of Parliament retired to Hanover and their Country Seats." According to Aubin, "something new and diverting would be welcome to the Town."[14] Her critique of the court's behavior addresses British anxiety regarding a foreign-born king on the English throne. Her strategic alliance mirrors her concern to defend a virtuous British identity threatened under a king (George I) who preferred not to reside in England. In the realm of fiction, Aubin constructed a model of Great Britain that allowed for the willful absence of its king when, depending on ones' allegiances, there were two kings of England "on the other side of the water" at the same time. Aubin also informs her readers that her booksellers had advised her

> to write rather more modishly, that is less like a Christian, and in a Style careless and loose, as the Custom of the present Age is to live. But I leave that to the other female Authors my Contemporaries, whose Lives and Writings have, I fear, too great a resemblance.[15]

In this context, Aubin's tales of a globalized Christian piety could be seen not only as a shaming of the selfish pleasures of the king and of amatory fiction writers, but also as a call to establish a morally reformed England. In her Mediterranean novel settings, Aubin dramatized her anxiety about the king's absence "beyond the Sea" by projecting it onto English travelers who leave home in order to gain riches through commercial networking with the Ottomans.

Aubin included extended accounts of European entanglements with Ottomans in the Mediterranean throughout her fiction, for instance in *The Strange Adventures of the Count de Vinevil and His Family* (1721), *The Noble Slaves* (1722), *The Life of Charlotta du Pont, an English Lady* (1723), and *The Life and Adventures of the Young Count Albertus* (1728), a continuation of her earlier novel *The Life and Adventures of the Lady Lucy* (1726). Rather than valorizing projectors like Crusoe, Aubin focused on the body of the noble male martyr—specifically her earliest hero, Count Vinevil, and her final hero, Count Albertus—and on the global Christian network that these human nodes helped weave through their altruistic examples.

Yet while her focus tends to be on the redemptive power of a noble man's martyrdom, Aubin also uses narratives of sexual violence to depict a martyrdom unique to women. The suffering of Aubin's fictional heroines adds another layer to the moral network that frames Vinevil's and Albertus's sacrifices. Two narratives of sexual violence shall demonstrate how Aubin employed female sacrifice of self, pleasure, and marital choice to inscribe her ideal of regained Christian unity into the plots of her novels.

Aubin's use of cultural hybridity to explore the violence and intimacy of empire is both unsettling and familiar. Describing John Dryden's difficulty in reconciling the sordid (but successful and therefore exemplary) history of Spanish conquest in the Americas with England's own imperial ambitions in *The Indian Emperour* (1665), Cynthia Lowenthal has coined the term "displacement" and argued that Dryden used this narrative strategy to present "native inhabitants" as possessing "free choice in the matter of their conquest."[16] The violent incursion of European imperialism thus registers instead as an "enlightened decision by the natives." As Daniel Goffman observes, the focus of early modern European trade shifted, between the seventeenth and eighteenth centuries, from the Mediterranean to "the Americas and East Asia."[17] But despite this perspectival shift, Barbary piracy in the Mediterranean continued to preoccupy the European imaginary.[18]

According to Eve Tavor Bannet, by 1800 the use of Barbary narratives as a template for slave narratives had become so conventional that "American writers were using white slavery in North Africa as a 'mask' for critiques of African slavery in the United States."[19] In other words, Lowenthal's "displacements" occurred not only between the Old World and the New, but also between the Mediterranean and the Atlantic and between the Americas and East Asia. Aubin's narratives of sexual violence introduce a network of intraliterary symbolic cross-references to depict the triumph of Christian unity over both the transatlantic slave trade and the perils of the Ottoman Mediterranean. By rejecting both slavery and the "Empire of the Eastern World," Aubin's characters were rejecting potential imperial models for Britain.

3

The drama of *The Strange Adventures of the Count de Vinevil and His Family*, for example, initially hinges on the lustful attentions of Mahomet, the Bassa of Constantinople's son, toward Ardelisa, a beautiful European Christian. But Ardelisa's vulnerability to Mahomet's power is framed in terms of her father's rashness, a failing for which the Count de Vinevil feels he must sacrifice his life. Vinevil leaves his native France for Constantinople after "finding his Estate impoverished by continued Taxations, and himself neglected by his Sovereign, and no ways advanced, whilst others less worthy were put into Places of Trust and Power."[20] Though Vinevil is a sympathetic character, this choice to forsake his homeland in order to go "amongst Mahometans, to avoid the seeing those, who have been my Vassals, lord it over me" is characterized as a rash and dangerous step by Longueville and Ardelisa and, ultimately, by Vinevil himself.[21] Vinevil's choice demonstrates a preference for worldly pleasure and greatness over resignation to Providence. This is signaled early in the novel when Vinevil tells Longueville, whom he considers as a son, that he does not want to expose him to the dangers he himself is going to encounter. Longueville, the novel's hero, chooses to remain enmeshed in an affective Christian network rather than to pursue worldly greatness. In a rhetorical question addressed to Vinevil he asks:

> Can you believe me capable of an Action so base, as to abandon you and Ardelisa, to whom my Soul is devoted, out of whose Presence I would not live, to gain the Empire of the Eastern World? No, my Father, your Fortune shall be mine; we will live and die together, nothing but Death shall ever separate us.[22]

Longueville chooses a unified Christian domestic arrangement over material allurements. He also secretly doubts that Vinevil has behaved prudently. He later warns Ardelisa that Vinevil's "Ambition and Resentments" will "cost both him and us" because Ardelisa will be particularly sexually threatened in Constantinople where "Religion shews itself in splendor . . . but . . . the vile Impostor's Name is echoed through the empty Quires and Vaults; where cursed Mahometans profane the sacred Piles, once consecrated to our Redeemer."[23] Longueville's allusion refers to Hagia Sophia, a Christian basilica converted to an imperial mosque after the Ottoman conquest of 1453. Just as the basilica dedicated to the holy spirit of wisdom[24] could be repurposed to perpetuate what Longueville considered to be the imposture of Islam, so Ardelisa's body might be taken over to reproduce Ottoman subjects and, perhaps, soldiers, perpetuating a military and trade dominance already confirmed by the Conquest of 1453. Ardelisa admits that she, too, did not "dare" to tell Vinevil that she had misgivings about the journey nor that she had felt a "Dread ever since we left our native Land."[25]

The characters' fears are shown to be well founded. Although Longueville warned Ardelisa to hide herself from public view as much as possible, Vinevil is delighted by his success in trading at Constantinople and "so many Bashaws and Persons of Quality came to [his house] to traffick for European Goods, that [Ardelisa] could not avoid being sometimes seen." Mahomet, just such a powerful and "lustful Turk" as Longueville had predicted, determines that he must have her; his sexual aggression confronts Vinevil with the full consequences of his choice to spend "the rest of his Days" among the pleasures of a foreign, non-Christian city where Aubin presents material wealth, like the conversion of Hagia Sophia, as symbolizing greatness rather than integrity. Vinevil, menaced by Mahomet, realizes that he has made a fatal error in exposing himself, Longueville, and Ardelisa to Mahomet's power and confesses: "too late I see my Rashness, for which I know you must condemn me."[26] Ultimately, he sacrifices himself so that Ardelisa and Longueville can escape, declaring that he prefers Ardelisa's honor to "Empires" before admitting: "It is just, my God, that I, who have exposed my Child, should first feel the Misery my Rashness merits, but do not let her perish here: Preserve her, Great Creator, from the Lust and Rage of these vile Infidels, and . . . let my Blood expiate all my Sins."[27] Ardelisa and Longueville experience many adventures and trials before they reunite, but the narrator concludes the novel by noting that "Divine Providence, whom they confided in, tried their Faith and Virtue with many Afflictions, and various Misfortunes; and, in the End, rewarded them according to their Merit."[28] Significantly, the narrator characterizes this fictional narrative as *proof* of divine Providence, of Longueville's "prophetick Fears," and of an imagined geography that associated Islam with material wealth, consumerist pleasure, religious intolerance, abuse of political power, and sexual predation. The ultimate union of Longueville and Ardelisa—enabled by Vinevil's repentance and expiatory martyrdom and by their trust in Providence—establishes a network of symbols that subtends Aubin's later novels, particularly her story of evangelism and martyrdom, *The Life and Adventures of the Young Count Albertus*. Like Vinevil, Albertus is a nobleman whose martyrdom serves as a moral resource for a Christian community turned away from worldly pleasure.

Aubin's novel *The Life and Adventures of the Lady Lucy* details the meeting of Count Albertus's parents against the backdrop of violence in Ireland following the Glorious Revolution. Albertus is the son of Lady Lucy, who is of a genteel Irish Catholic family; her father is killed when William of Orange invades Ireland. Lucy and her mother are saved from ill treatment at the hands of the invading soldiers by throwing themselves on the mercy of the "German captain," Lewis Augustus Albertus, who later marries Lucy.[29] I will pass over a rather complicated plotline in which Lewis Augustus becomes convinced that his innocent wife is pregnant with another man's child and tries to assassinate her. Lucy miraculously survives, unbeknownst to her husband, and safely delivers a son. The young Count Albertus is thus the Catholic son of an aristocratic Irish Stuart Royalist mother and an aristocratic German father employed by the Dutch King William. Though violently assaulted by a deceived Williamite, Albertus's mother continues to love and pray for her husband and welcomes him when he returns to her as a penitent. The young Count Albertus therefore literally embodies, in his aristocratic bloodlines, the union of Jacobite and Williamite allegiances tied together by Christian faith and providential benevolence. In the person of Count Albertus, Aubin unified Great Britain at a time when the country was still divided over the legitimacy of the Stuart and Hanoverian kings.

As an episodic novel that combines amatory exploits, moral didacticism, and travel narrative, *The Life and Adventures of the Young Count Albertus* focuses on the adventures of the title character after he becomes a monk and decides to journey to China as a missionary. Albertus, like Vinevil, is a nobleman. His status is important because it registers the extent of his sacrifice (of material goods, prestige, privilege, eventually of his life) in becoming a monk, undertaking an evangelical mission to China, forming a community of Christians in North Africa when he is shipwrecked, and eventually submitting to martyrdom in Tartary. In both *The Life and Adventures of the Lady Lucy* and *The Life and Adventures of the Young Count Albertus* Aubin experiments with a complex cultural hybridity in order to reconcile divisions within Britain. She does this by hybridizing and essentializing noble blood and by dislocating her noble character from his domestic environment through evangelism. He thus becomes a node in a global Christian network held together by his example and, ultimately, connected by his blood. Aubin specifically privileged royal or aristocratic conversion as a means to restore Christian unity because a nobleman's exemplary behavior would most effectively illustrate the immensity of a sacrifice of worldly pleasures in exchange for an ascetic life that leads to heavenly glory.

Yet another type of networking comes into play when his wife is dying. Tired of the world, Count Albertus decides to become a monk and journey to China "with no other View but the Conversion of Pagans and Infidels to Christianity," thereby becoming a member of a worldwide network of evangelism.[30] As happens frequently in Aubin's fiction, his ship is wrecked near Tunis. Though initially expecting to be captured, enslaved, or put up for ransom, he is aided by a Frenchwoman, a former slave and convert to Islam, who takes him to an old mosque and disguises him as "Santoin" or "Mahometan Religious."[31] In this disguise, he encounters many Christians entangled, like himself, in the network of Ottoman imperial commerce and desperately trying to return to Europe. This scenario enables Aubin to depict the related and overlapping experiences of suffering Christians, both men and women, as they respond to temptations and trials.

Conversion to Christianity is paramount for Aubin and facilitates a variety of other interpersonal constellations characterized by hybridity, all of which turn female cultural interlopers into human nodes within a global network of martyrs. The character of Anna in *The Life and Adventures of the Young Count Albertus* is an example. After Albertus has gathered about him a small community of noble Christian escapees from Ottoman or Moorish violence, he meets the noble Moor Abra and Anna, the European woman Abra has forced to live in a cave with him for five years. After Abra describes how he stole Anna away from the emperor and forced her to remain sequestered with him and to bear his children, he informs Albertus that he would convert to Christianity if Anna were to marry him. Faced with opportunity to convert a princely Moor, Albertus shrugs off Anna's numerous objections. Abra is a royal and ready to convert; his conversion takes precedence over everything else, including the trauma of a female Christian rape victim. As Albertus tells Anna:

> Well, Madam, why are you thus afflicted? Weep no more; see
> Heaven here sends you Christian Friends, your Life shall be no
> longer wretched; we can provide you a Companion, a Lady
> nobly born, as we believe you are; rejoice that the Almighty has
> by your Means drawn this Royal Infidel to be a Christian, you
> must now resolve to make him your lawful Husband, since
> Providence had given you to him.[32]

For Anna, the marriage is an emotional martyrdom that requires her to sacrifice her love for another man. This transcultural marriage reinforces the primacy of Christian unity over every other consideration, including those of sexual and domestic violence. Further, Anna's marital martyrdom is a foretaste of the violence of Augustus's death.

Abra is ultimately baptized as Bartholomew, "it being the Festival of that holy Apostle."[33] The date is historically significant. The Saint Bartholomew's Day Massacre of French Huguenots, August 23–24, 1572, was one of the most notorious acts of violence that Catholics committed against Protestants in the early modern period.[34] The significance of Abra's Christian name corresponds with Aubin's consistent justification and belittlement of transnational violence in the interest of Christian domestic unity. The male perpetrator is redeemed through the act of conversion and the birth of Christian children of mixed ethnic heritage. Abra's and Anna's children are—as Count Albertus himself is—the evidence that Christian unity is possible even within a context of transnational violence.

Significantly, a similar scene in *The Life of Charlotta du Pont, an English Lady* (1723) suggests that Aubin viewed profiting from the transatlantic slave trade to be as reprehensible as trading with the Ottoman Empire. After the eponymous heroine is stranded on an island somewhere in the Caribbean, she and her companions stumble upon a couple—Domingo, an Angolan prince and ex-slave, and Isabinda, the daughter of the Virginia planter who had owned Domingo. Together these two characters have had a "Molotto child."[35] Isabinda describes to Charlotta how Domingo kidnapped and raped her and has kept her a prisoner on the island for two years. Yet this scene, too, subordinates transnational violence to the great goal of regained Christian unity. While Aubin's privileging of white skin is obvious, her central distinction is a moral, not a racial one: Life in a Christian society is presented as incompatible with the violence of the slave trade. Though she at first "wept for Joy to see a Christian white woman" and while she believes that Domingo would "certainly kill" her if her father tried to take her away from him, Isabinda assures Charlotta that Domingo is a Christian and that she would marry him if they could only leave "such a solitary miserable Life" as they have lived on the island. Charlotta promises her aid, believing that Isabinda now loves and has forgiven Domingo. Charlotta exclaims that the

> selling human Creatures is a Crime my Soul abhors; and Wealth so got never thrives. Tho' he is black, yet the Almighty made him as well as us, and Christianity ne'er taught us Cruelty: We ought to visit those Countries to convert, not buy our Fellow-Creatures, to enslave and use them as if we were Devils, or they not Men.[36]

This scathing indictment of the slave trade shows that Aubin's model of hybridity included racial others, while consistently excluding non-Christian others. Just like the pleasures and luxuries of Ottoman trade, Aubin depicts the commercial network of the transatlantic slave trade as antithetical to Christian unity.

Beyond the similarities of the Anna and Isabinda episodes, several of Aubin's novels include nested hermit episodes that show men, formerly involved in commercial networks and dispersed along their trade routes, who have withdrawn from worldly pleasures and become nodes in a global network of piety and asceticism. In Aubin's fiction these hermits populate the woods around Tunis, welcoming and protecting escapees from prison and the seraglio and enabling them to connect in a seemingly hostile wilderness. For instance, Count Albertus at one point saves a man who turns out to be Don Gomez (the father of one of Albertus's companions), who had previously been sheltered by an English hermit who in turn had himself been rescued by an Italian hermit, stranded near Tunis on his way to do missionary work in Japan. This strange coincidence creates a parallel between Count Albertus and the nameless Italian as both had intended to journey to East Asia on missions as members of a network of evangelism.

At this juncture, the narrator offers an extended commentary, contrasting the virtuous example of these evangelical hermits and the corrupt pleasures of contemporary European Christians. The narrator reflects on the

> profuse Manner in which Christians live in Europe, and how
> unconcerned great and rich Men sit down to their Tables . . .
> whilst thousands of poor Christians perish in loathsome Prisons
> and . . . are left in slavery in the cruel Hands of Infidels who
> delight in torturing and tormenting their poor Fellow-
> Creatures, particularly such who will not forsake their savior for
> the Imposter Mahomet.[37]

Using the nameless hermit as an example, the narrator points out that it is possible to prefer austerities to all the pleasures that

"Worldlings" relish. The criticism of European luxury is underscored by the fact that the nameless hermit refuses to return to Europe with the other Christians when they find an opportunity to escape. After they return to Europe, Count Albertus stays put for only three more years before setting out on another mission to China. By giving the characters of the Italian hermit and Count Albertus the free choice to remain in or return to the East, Aubin reiterates her main moral lesson: There can be no redemption before luxury trade goods have not been replaced by Christian martyrial blood. The commercial network of the Silk Road must make way for the alternate network of Christian altruism. Count Albertus has to return to the East, where his noble blood will become the providential commodity in a symbolic act of self-sacrifice, before beginning the return journey along the Silk Road to a disunited England enfeebled by luxury. Just as the sun rises in the East, so Aubin envisions the son of united but dislocated Britain rising through martyrdom, the rays of his example spreading from East to West through the medium of oral and print narratives—just as Aubin hoped her exemplary tales would do.

Before this pivotal event occurs, Albertus is welcomed in the "great Empire" of China, learns the local language, and makes many converts before being captured by "Tartars" and sent to "Turquestan." There he becomes tutor to the general's son, converts him to Christianity as well, and is eventually martyred. Commenting on the martyrdom of Albertus, the narrator momentarily steps out of the narrative to incite the English to prefer martyrdom to pleasure: Albertus ran

> the Hazards of a long and dangerous Voyage to convert Infidels and Pagans to Christianity, and gain that Crown of Martyrdom which so few in this unthinking Age do court or endeavor to obtain, sealing the Truth of the Doctrine he had taught with his Blood.

The intention is to "do a just Honour to the Memory of so excellent a Man, and with Intent to excite others to follow so holy and brave an Example." But Aubin's narrator issues a backhanded challenge to English readers, too, disdainfully admitting that

> I forgot that I am speaking in a Nation and to a People who are the greatest Part of them more fond of Pleasure than Martyrdom, and care not to be reminded of Death; yet I hope there is a great Number of good Christians amongst us who are truly zealous for God and Religion, and would not scruple to lose their Lives and Fortunes in a good Cause: These I honour, and to these I dedicate this History, hoping they will . . . admit me into the Number of their Friends.[38]

Aubin situates herself and her fiction within the same affective moral network as her noble martyrs. Wandering in a wilderness of corrupt pleasure, the novelist cajoled, shamed, and endeavored to persuade readers to live like her characters. In her fiction, Aubin transformed the "Empire of the Eastern World" from the original distribution center for luxury goods within a worldwide network of commerce into the true north of Christian identity, exporting that most precious of Christian treasures, the blood of a martyr.

Count Albertus's blood not only unifies Jacobite and Williamite England, but also China, Turkestan, the Ottoman Empire, and all of Europe: his martyrdom contributes to a new network—a new Silk Road—facilitating the distribution of the alternate commodity of moral example rather than the material pleasures of trade. Ironically, Aubin portrays an evangelical mission to the Far East as the only way for British identity to reorient itself toward self-sacrifice and to reject the materialism that has previously undermined Christian integrity. The new identity depends on an imagined geography of a predatory Muslim Mediterranean and a wealthy and welcoming Far East that was at least partly vulnerable to brutal Muslim invasion. Ultimately—and despite her unique hybrid model of British Christian identity—Aubin's alternate Silk Road projected England as the unified center of an imagined non-Christian network, deprived of nuance, complexity, and its own moral integrity.

Notes

1. "network, n. and adj.," *Oxford English Dictionary*, 2nd ed. (Oxford: Oxford University Press, 1989), online edition, http://www.oed.com/. See definitions 1, 2a, 2b, and 3.

2. Penelope Aubin, *A Collection of Entertaining Histories and Novels, Designed to Promote the Cause of Virtue and Honour* (London: Printed for D. Midwinter, 1739), 2:iii–iv. All quotations from Aubin's works are taken from this collection. I have silently modernized all instances of the long "s" and the use of italic script in the prefaces.

3. Dwight Codr, *Raving at Usurers: Anti-Finance and the Ethics of Uncertainty in England, 1690–1750* (Charlottesville and London: University of Virginia Press, 2016), 1–2.

4. Codr, *Raving*, 2.

5. Eve Tavor Bannet, "Captivity and Antislavery," chap. 2 in *Transatlantic Stories and the History of Reading, 1720–1810: Migrant Fictions* (Cambridge: Cambridge University Press, 2011), 48.

6. Codr, *Raving*, 109.

7. Susan Staves, *A Literary History of Women's Writing in Britain, 1660–1789* (Cambridge: Cambridge University Press, 2010), 194, has already characterized Aubin's fundamentally conservative theological vision as "we should trust in God to relieve our miseries," since if He "elects to have us suffer in this world, he will relieve us in the next."

8. Debbie Welham, "The Particular Case of Penelope Aubin," *Journal for Eighteenth-Century Studies* 31, no. 1 (2008), 69; Debbie Welham, "The Political Afterlife of Resentment in Penelope Aubin's *The Life and Amorous Adventures of Lucinda* (1721)," *Women's Writing* 20, no. 1 (February 2013), 49–63.

9. Sarah Prescott, "Penelope Aubin and *The Doctrine of Morality*: A Reassessment of the Pious Woman Novelist," *Women's Writing* 1, no. 1 (1994), 99–112; Aparna Gollapudi, "Virtuous Voyages in Penelope Aubin's Fiction," *Studies in English Literature, 1500–1900* 45, no. 3 (Summer 2005), 670; Edward Kozaczka, "Penelope Aubin and Narratives of Empire," *Eighteenth-Century Fiction* 25, no. 1 (Fall 2012), 199; Chris Mounsey, "'. . . bring her naked from her Bed, that I may ravish her before the Dotard's face, and then send his Soul to Hell':

Penelope Aubin, Impious Pietist, Humourist or Purveyor of Juvenile Fantasy?" *Journal for Eighteenth-Century Studies* 26 (2003), 59; Chris Mounsey, "Conversion Panic, Circumcision, and Sexual Anxiety: Penelope Aubin's Queer Writing," in *Queer People: Negotiations and Expressions of Homosexuality, 1700–1800*, ed. Chris Mounsey and Caroline Gonda (Lewisburg, PA: Bucknell University Press, 2007), 246–260; Bannet, "Captivity and Antislavery," 54–55; Adam R. Beach, "Aubin's *The Noble Slaves*, Montagu's Spanish Lady, and English Feminist Writing about Sexual Slavery in the Ottoman World," *Eighteenth-Century Fiction* 29, no. 4 (Summer 2017), 583–606.

10. Aubin, *Count de Vinevil*, in *Collection*, 2:4.

11. Aubin, *Charlotta du Pont*, in *Collection*, 3:140–141.

12. Aubin, *Collection*, 3:iii, vi. Aubin addresses the letter to "Mrs. Rowe." This was long assumed to be Elizabeth Singer Rowe, but Chris Mounsey has pointed out that this could not be the case since "Mrs. Rowe" is identified as the daughter of the Dean of Exeter ("'. . . bring her naked,'" 61–62).

13. Aubin, "Preface," *Collection*, vol. 1, unpaginated. Jane Spencer suggests that the anonymous writer of the preface might have been Samuel Richardson. Jane Spencer, *The Rise of the Woman Novelist: From Aphra Behn to Jane Austen* (Oxford: Basil Blackwell, 1986), 86–88.

14. Aubin, prefatory letter, *Collection*, 3:vi.

15. Aubin, prefatory letter, *Collection*, 3:vi.

16. Cynthia Lowenthal, *Performing Identities on the Restoration Stage* (Carbondale and Edwardsville: Southern Illinois University Press, 2003), 42.

17. Daniel Goffman, *The Ottoman Empire and Early Modern Europe* (Cambridge: Cambridge University Press, 2002, 2009), 233.

18. Bannet, "Captivity and Antislavery," 49, points out that Aubin's novel *The Noble Slaves* roughly coincided with the highly publicized release of "260 British slaves" by the emperor of Morocco. Later in the century, Elizabeth Griffith would re-edit *The Noble Slaves* (1777), framing it in abolitionist terms. After crossing the Atlantic, the novel also came to be used in American abolitionist arguments.

19. Bannet, "Captivity and Antislavery," 60.

20. Aubin, *Count de Vinevil*, in *Collection*, 2:1.

21. Aubin, *Count de Vinevil*, in *Collection*, 2:3.

22. Aubin, *Count de Vinevil*, in *Collection*, 2:4.

23. Aubin, *Count de Vinevil*, in *Collection*, 2:6, 5.

24. For the association of Jesus Christ with Holy Wisdom (*Hagía Sophía*) in the Eastern Orthodox and the Catholic churches and for the dedication of church buildings (such as the basilica Hagia Sophia in Constantinople) to Christ as the personification of Divine Wisdom, see Gerald O'Collins, *Christology: A Biblical, Historical, and Systematic Study of Jesus* (Oxford: Oxford University Press, 2009), 35–41.

25. Aubin, *Count de Vinevil*, in *Collection*, 2:6.

26. Aubin, *Count de Vinevil*, in *Collection*, 2:10.

27. Aubin, *Count de Vinevil*, in *Collection*, 2:16.

28. Aubin, *Count de Vinevil*, in *Collection*, 2:73.

29. Aubin, *Lady Lucy*, in *Collection*, 2:83.

30. Aubin, *Albertus*, in *Collection*, 2:237.

31. Aubin, *Albertus*, in *Collection*, 2:238.

32. Aubin, *Albertus*, in *Collection*, 2:264.

33. Aubin, *Albertus*, in *Collection*, 2:267.

34. See John Marshall, *John Locke, Toleration and Early Enlightenment Culture* (Cambridge: Cambridge University Press, 2009), particularly chap. 2, "Catholic Intolerance and the Significance of Its Representations in England, Ireland, and the Netherlands, c. 1687–92," 55–93, at 58. Further, Pope Gregory XIII "designated 11 September 1572 as a joint commemoration of [the 1571 defeat of the Ottomans at] the Battle of Lepanto and the massacre of the Huguenots." See Eunice Howe, "Architecture in Vasari's 'Massacre of the Huguenots,'" *Journal of the Warburg and Courtauld Institutes* 39 (1976), 258.

35. Aubin, *Charlotta du Pont*, in *Collection*, 3:63.

36. Aubin, *Charlotta du Pont*, in *Collection*, 3:71.

37. Aubin, *Albertus*, in *Collection*, 2:291.

38. Aubin, *Albertus*, in *Collection*, 2:307.

5

Robert Morrison and the Dialogic Representation of Imperial China

JENNIFER L. HARGRAVE

1

History often casts nineteenth-century European missionaries as complicit in imperial expansion.[1] Within the British Empire, missionaries' conversion of local populations not only fulfilled a perceived moral responsibility but also satisfied the imperial need for governable colonial subjects. As James L. Hevia aptly observes, nineteenth-century imperialism relied upon "the production of knowledge about indigenous peoples and their social practices, knowledge that could then be deployed to manage, monitor, and reorganize populations."[2] Missionaries often contributed to the British imperial acquisition and generation of knowledge through the study of local religious customs and the translation of religious doctrine. Accordingly, early nineteenth-century missionaries to China, alongside diplomats and East India Company employees, were responsible for molding Britons' understanding of the Far East. According to Peter J. Kitson, "[t]heir publications, journals, and libraries became the foundation of the institutionalized study of China in Britain as well as of more popular understandings."[3] Although British missionaries often produced new knowledge of

China for an eager British readership, their publications did not always complement or conform to emerging imperial ideologies.

The first British Protestant mission to China comprised one individual, Robert Morrison (1782–1834). Sponsored by the London Missionary Society, Morrison arrived in Canton (today's Guangzhou) in 1807. The London Missionary Society instructed Morrison, during his initial years in China, "to concentrate on learning Chinese and preparing written materials that would be of continuing value in a long-range evangelization effort."[4] Morrison's influence upon Protestant missions to China was very much "long-range" and assumed more of an "iconic importance," argues Kingsley Bolton, "for Christian missionaries, Western linguists, sinologists, educationalists, and the treaty-port communities of late-Qing and Republican China."[5] During his lifetime, however, Morrison failed either to instigate widespread Chinese conversion or to augment British imperial influence in the Far East. Instead, the contemporary reception of Morrison's work, though laudatory, focused on his scholarly accomplishments rather than his evangelism.[6] Through his Chinese-English-Chinese translations, his meticulous sinological studies, and his cofounding of the Anglo-Chinese College in Malacca, Morrison cultivated British intellectual interest in Chinese history, language, and culture. Famed diplomat and sinologist George Thomas Staunton honored Morrison by deeming him the "first Chinese scholar in Europe."[7] In 1824, Morrison's scholarship on China—rather than his missionary work—was applauded publicly through his induction into the Royal Society.

Nevertheless, modern scholarship perpetuates the image of Morrison as the father of Protestantism in China and, therefore, as complicit in British imperial expansion. Scholars often derive their understanding of Morrison's legacy from his second wife Eliza Morrison's *Memoirs of the Life and Labours of Robert Morrison* (1839). In one of the few comprehensive studies of Morrison's scholarship, Christopher Daily asserts that Eliza Morrison's *Memoirs* "effectively inflated the Christian public appreciation for her husband's mission by painting him in a virtuous light . . . as a triumphant and creative pioneer."[8] There is much to be gained by accepting Eliza Morrison's hagiographical depiction of her husband. By focusing almost entirely on Morrison's English-Chinese and Chinese-English translations of religious texts (e.g., the Bible and Confucian teachings) and the influence of these translations on later nineteenth-century Protestant missions, scholars from the nineteenth century forward create a portrait of Morrison that not only confirms his enthusiastic evangelism but also establishes the missionary as an active contributor to Britain's imperialist agenda in the Far East. Suzanne W. Barnett, for example, narrates a unilateral transfer of knowledge from British missionaries to the Chinese, knowledge that would "provide information about Western civilization and culture that would encourage Chinese acceptance of Christianity."[9] In short, even the introduction of secular Western knowledge contributed to global evangelism and eased imperial expansion. The modern Chinese government sustains this imperialist account of Morrison by subsuming his work within what Shu-mei Shih characterizes as, "the nationalist historiography of modern China" in which "Chinese history since the Qing is nothing but a series of opium wars, unequal treaties, Western imperial acts of aggression, and China's relentless search for sovereignty under duress."[10] Within this context, the Chinese government continues to link Morrison with British imperialism—even the illegal opium trade—despite evidence to the contrary.[11]

Significantly, these accounts of Morrison's relationship to China elide his explicitly didactic texts: a language primer, *Dialogues and Detached Sentences in the Chinese Language* (1816), and a children's textbook, *China; A Dialogue, For the Use of Schools* (1824).[12] It seems incontrovertible that Morrison's translated religious texts would prove foundational to the establishment of Protestant missions later in the nineteenth century. But by focusing exclusively on Morrison's sacred scholarship, scholars force his representations of early nineteenth-century Anglo-Sino exchanges—whether economic, diplomatic, religious, or intellectual—to cohere to a teleological account of British imperial history that culminates in the mid-century Opium Wars. Morrison's educational texts convey a different story. They reflect Romantic-era Britons' more ambiguous attitudes toward the Chinese.

In the decades preceding the first Opium War (1839–1842), British exchanges with the Chinese emerged from a position of intellectual curiosity as well as imperial self-interest. Early nineteenth-century Britons—such as statesman John Barrow, linguist William Marsden, and even poet William Wordsworth—acknowledged that China was a self-sufficient, culturally sophisticated empire while simultaneously developing an imperial desire for its subjugation. Morrison's *Dialogues and Detached Sentences* and *China; A Dialogue* are illustrative examples that reflect the missionary's attempt to develop mutually informative relationships with Chinese individuals that were not premised on conversion or grounded in imperial ideology. Instead, Morrison depicts the potential for foreign relations to foster cultural exchange, rather than the unidirectional transfer of knowledge. Rather than assuming British cultural and imperial superiority, Morrison's seemingly simple didactic texts depict a complex and multifaceted China that is admirable yet flawed, civilized yet immoral, global yet isolated. This ambiguous portrayal of China instills Morrison's readers with neither the imperialist nor the religious fervor to subjugate China. Instead, he encourages Britons, especially British youth, to study and learn from the Chinese, a position that dissociates him from both missionaries residing elsewhere in the British Empire and the more typically rapacious imperialists employed by the East India Company.

In the following pages, I argue that Morrison's didactic texts exhibit a combination of admiration for and criticism of Chinese culture that challenges modern scholars' depiction of nineteenth-century representations of British contact zones as unequivocally imperialist. In the latter half of the eighteenth century, travel literature emerged as the predominant mode of representing Britain's increasing number of global contact zones and of educating Britons about their expanding empire. Mary Louise Pratt identifies sentimental travel writing, writing whose "[a]uthority lies in the authenticity of somebody's felt experience," as being particularly instrumental in the civilizing missions that rationalized British imperial expansion throughout the nineteenth century.[13] For example, in her reading of Mungo Park's *Travels in the Interior of Africa* (1799), Pratt acknowledges the Scottish explorer's affirmation of "African agency and experience," while also noting that his assessments of African culture never directly challenge "European authenticity, power, and legitimacy."[14] For Pratt, travel literature is a "strateg[y] of representation whereby European bourgeois subjects seek to secure their innocence in the same moment as they assert European hegemony."[15] Implicit in Pratt's description of travel literature is the unidirectional application of knowledge. Britons produced knowledge of foreign spaces less for their own edification and more as a means of securing colonial governance. While I do not question the cogency of Pratt's argument regarding travel literature, this essay explores alternative textual representations of Britain's contact zones. The Morrison texts under examination here rely almost entirely, as their titles indicate, upon dialogues. I argue that the "dialogue" offers a rhetorical means of escaping the Romantic-era expectations held of colonial literature by allowing the author to assume multiple subject positions and, consequently, to imagine the Anglo-Sino contact zone from diverse, even contradictory, positions. Via these didactic dialogues, Morrison advocates a neutral approach to China that observes, questions, and reflects upon cultural differences without advocating an immediate recourse to imperial thought or action.

2

Throughout his career in China, Morrison found himself in a precarious position. According to Qing government code, Europeans were forbidden from instructing the Chinese in Christian doctrine or from printing Chinese translations of sacred texts for the purposes of proselytism.[16] Accordingly, Morrison needed a position that would justify his residence in Canton and would supplement the London Missionary Society's meager financial support. As Morrison's Chinese language skills developed, he quickly found himself of value to the East India Company. By 1809, Morrison held allegiance to not only the London Missionary Society but also the East India Company by whom he was employed intermittently as a translator until the end of the company's monopoly in 1834. Morrison's position with the company led to his eventual appointment as interpreter for the second British embassy to China led by Lord Amherst in 1816. Because of his dual entanglements with the London Missionary Society and the East India Company, modern scholars logically conclude that Morrison held imperialist sympathies. Chris Murray, for example, argues that Morrison's Chinese-English translation of Confucius allowed him to further his mission indirectly without posing a threat to British commercial interests: "Confucian works related to Christian activity in a manner that appealed to . . . Morrison's financial supporters, but were sufficiently far-removed from overt proselytizing to allay the suspicions of administrators in Asia."[17] In addition, studies of Confucius provided new insight into the mindset of China's ruling elite, insight that the East India Company required to leverage their relationship with the Cantonese customs administrator, or *Hoppo*.[18] Morrison himself publicly acknowledged his financial dependence upon the company to publish his work.[19]

Yet Morrison also bore a third allegiance to his experiential knowledge of China, which diverged from the attitudes adopted and promoted by both the London Missionary Society and the East India Company. Despite his value as a translator, the company threatened, more than once, to terminate Morrison's employment due to his continued (though often derailed) efforts to proselytize as well as his more scholarly approach to Chinese culture, an approach eventually made manifest in his cofounding of the Anglo-Chinese College with William Milne in 1818. Similarly, the London Missionary Society increasingly perceived Morrison's Chinese studies as diluting his evangelical commitments and, therefore, decreased their financial support of his missions throughout the last decade of his life. Morrison's multiple, and constantly fluctuating, investments in China and its culture make an unequivocal reading of *Dialogues and Detached Sentences in the Chinese Language* unfeasible.

Published nine years after Morrison's arrival in China, *Dialogues and Detached Sentences* makes manifest his willingness to communicate and engage with Chinese locals. His text is a collaborative effort: many of the dialogues and their internal stories were provided to Morrison by his Chinese tutors and acquaintances. The topic selection for each dialogue—which include "With a Tea Merchant," "With a Teacher of Chinese," and "Respecting the Seizure of a Thief"—testifies to Morrison's careful study of Chinese social expectations and their impact on cross-cultural exchanges between Cantonese and English merchants. Moreover, these crafted dialogues refrain from expressing any cultural or nationalist allegiance. For example, the opening dialogue between a Chinese merchant and a potential customer focuses on the shop's lack of merchandise because "the European ships have not yet arrived."[20] While the late arrival of foreign merchandise would be of concern to Canton merchants, the Eurocentric nature of this dialogue suggests Chinese dependence upon European trade. Such a sentiment would align Morrison with prevalent company attitudes. However, the dialogue also exhibits subtle critiques of European affairs. The European ships' late arrival, explains the merchant, may be attributed to "an unsettled state at sea; that the foreigners are always fighting, and hence they arrive late."[21] Morrison, in this seemingly innocuous line, alludes to seventeenth- and eighteenth-century perceptions of China as the pinnacle of political stability.[22] By subtly mocking the geopolitical tensions that Europeans must navigate, the merchant suggests that the Chinese are relatively free of these problems and challenges. It is this ambivalent depiction of Chinese culture that typifies Morrison's early didactic work.

Regardless of its vacillating perspectives, Morrison's *Dialogues and Detached Sentences* provides valuable snapshots of contemporary Chinese culture that, though ostensibly peculiar by Western standards, are nonetheless presented without authorial critique. In his thirty-one dialogues and narratives, Morrison provides not only the vocabulary and idiomatic phrases that a Canton-based British traveler would find necessary but also the cultural knowledge needed to negotiate cross-cultural exchanges—economic, diplomatic, and philosophical. For British merchants, Morrison provides dialogues that enumerate the types and quality of Chinese tea or demonstrate the art of bargaining with Chinese tradesmen. For British diplomats, he depicts the Chinese courtesies expected of visitors upon entering either a Mandarin's home or the Chinese imperial court. For the British missionary, he anticipates an initial conversation with a Chinese local regarding the key attributes of China's three dominant religious sects—Buddhism, Taoism, and Confucianism. Some of the most insightful dialogues, however, are those in which Morrison addresses the general Chinese language learner.

In the dialogue "With an Assistant in Learning the Language," Morrison stages a conversation wherein a Chinese tutor recommends the books most instructive to the beginning student. Unlike most sinologists' dependence upon ancient Chinese literature for cultural insight, Morrison recognizes the importance of both ancient and contemporary Chinese texts to a European education in Chinese language and culture. To this end, his *Dialogues and Detached Sentences* recommends that the student of Chinese consult not only classic texts such as the *Da Xue* but also Cao Xueqin's *Hong Lou Meng* (in English, *Dream of the Red Chamber* or *The Story of the Stone*).[23] The *Da Xue*, one of the classic Chinese texts known as the *Si Shu* (in English, Four Books), was composed by Confucius and his student Zeng Zi prior to the third century B.C.E. and forms the foundation of Confucian thought. The *Si Shu* were considered mandatory reading for Chinese intending to take the imperial examinations. *The Story of the Stone*, though regarded today as a Chinese classic, could not be more different from the *Da Xue*. First published in 1792, with manuscripts circulating in the early 1760s, *Story of the Stone* follows the rise and fall of the Jia clan in early eighteenth-century Beijing. The novel quickly became the most popular Chinese novel of the eighteenth and nineteenth centuries and continues to be regarded as a source of great insight into eighteenth-century Chinese culture. However, until David Hawkes's groundbreaking 1970s translation, *The Story of the Stone* remained unavailable to English readers. Therefore, Morrison's citation of this text holds twofold importance. First, it further attests to Morrison's thorough study and appreciation of contemporary Chinese culture. Nowhere else in Romantic-era sinology is *Story of the Stone*'s cultural importance acknowledged. Morrison presciently recognizes the text's importance within Chinese culture and asserts its relevance to a European's understanding of the culture. Asserting a contemporary text's relevance to European sinology, however, was itself a controversial act and contributes to the secondary significance of Morrison's reference. Britons' assumption of cultural superiority was premised upon the Enlightenment notion of stadial theory, which established European societies as the pinnacle of civilized development and "oriental" nations as being culturally stagnant.[24] In expressing appreciation of contemporary Chinese literature, Morrison undermines imperialist claims that China was an empire in cultural decline.

3

Dialogues and Detached Sentences focuses on illuminating Chinese customs and teaching practical language skills that would facilitate Anglo-Sino exchanges in Canton. While *China; A Dialogue* continues to provide insight into Chinese culture, it also assumes a less practical purpose. Instead of addressing British merchants, diplomats, and missionaries, Morrison appeals to a domestic British audience less invested, for instance, in the minutiae of Chinese linguistics. Accordingly, he stages a series of ten conversations between a father—presumably Morrison—and his two children, Mary Rebecca and John. The premise is that these conversations occurred during a ship voyage from England to China. The dialogues' topics range from detailed descriptions of each Chinese dynasty to accounts of brutal Chinese punishments, from observations of

Buddhist monks' devotional practices to stories of infamous Chinese emperors. At the heart of Morrison's text is a nuanced explanation of Chinese religions vis-à-vis Christianity. However, the religious dialogues are delayed in their appearance by being sandwiched between less contentious topics. Often, Morrison describes elements of Chinese history or culture via a parallel aspect of European history or culture. These comparative moments do not affirm European superiority. Instead, they provide Morrison the leverage to develop a rational critique that deconstructs and disproves several stereotypes of Chinese culture.

Early nineteenth-century travel narratives often exhibit Britons' morbid fascination with female foot-binding. In his popular travelogue *Travels in China* (1804), John Barrow perceives the custom as "unnatural and inhuman," and yet concedes that "[f]ew savage tribes are without the unnatural custom of maiming," a statement that summarily degrades Chinese civilization.[25] Conversely, Morrison's discussion of female foot-binding counters British narratives that understand the custom as indicative of China's primitive culture. Curious about the lives of Chinese women, Mary Rebecca initiates a discussion of the custom and quickly labels the practice "a very foolish fashion" (*China; A Dialogue*, 110). Morrison agrees with her general critique but moves to nuance and mitigate her censure. First, Morrison establishes foot-binding as a cultural tradition with origins in the ninth century, rather than a fleeting fashion. In fact, Morrison makes this point more than once. While foot-binding's longstanding practice does not diminish its violence or condone its perpetuation, it does establish Chinese respect and affinity for tradition. Morrison further, and more convincingly, attenuates foot-binding's severity by drawing attention to equivalent British customs. He responds to his daughter's incredulity by asking, "Do English ladies never hurt themselves for fashion's sake . . . ? In very cold weather don't they wear very light dresses, because it is the fashion to do so; and by doing so, often injure their health. Formerly too the very tight stays used to hurt them; not very unlike the Chinese little shoe compressing the foot" (110). Morrison here uses British customs both to soften popular perceptions of China and to launch a minor criticism of Britain. Morrison's son responds to this latter point by accusing his father of "always blaming the English" (111). "Blame" is a curious word choice given the topic of conversation and belies the importance of this exchange to the work as a whole. Morrison defends his position by critiquing the two cultures equally: "I blame both the Chinese and English for being the slaves of fashion; and for doing what they think hurtful to their health because it is fashionable" (111). Morrison's scrutiny of a seemingly mundane issue—fashion—demonstrates how knowledge of China may be used not merely to bolster Britons' sense of superiority—made evident in Morrison's son's accusation—but rather to reveal British shortcomings.

Given his continued association with the London Missionary Society, it should come as no surprise that Morrison dedicates considerable space to discussing Chinese religion. What is surprising, however, is how his discussion does not revolve around affirming China as heathen. Instead, akin to his discussion of foot-binding, Morrison makes a more nuanced argument that simultaneously evacuates China of religion and affirms its cultural rectitude. Morrison recognizes China's three teachings—Confucianism, Buddhism, and Taoism; he focuses on the oldest and most prevalent of these teachings, Confucianism. Before delving into Confucianism's doctrinal particulars, Morrison reclassifies Confucianism as a moral philosophy rather than a religion. By dissociating Confucianism from religion, Morrison shapes his representation of China in three ways. First, he establishes a legitimate need for a Christian missionary presence since China ostensibly lacks religious guidance. Second, though in rather sharp contrast to his first purpose, Morrison applauds Confucian teachings without perceiving his appreciation as a challenge to Christian supremacy. In fact, he even admits to a number of similarities across Confucian and Christian doctrines. It is these similarities that allow Morrison to reverse his critical gaze onto England, with a particular focus on what he perceives to be its religious inadequacies. In essence, a dialogue that commences with Confucianism's demotion from ancient religion to philosophy concludes with a harsh judgment of Britons' religious failings.

It is important to remember that these ideas are conveyed through conversations with children, not among theologically astute missionaries. The simplicity of Morrison's presentation of these ideas belies their complexity, thereby making his argument more compelling. When teaching his children about Confucius, Morrison begins with the statement, "Confucius taught Ethics, rather than religion: he was what is called, a *moral Philosopher*" (70). This declaration prompts Mary to ask her father to distinguish between these terms. For Morrison, moral philosophy outlines human beings' duties to one another whereas religion stipulates their duties to God. This distinction situates Confucianism outside religion's purview, thereby justifying the missionary's role "to teach people *their duties to God, and the way of salvation*" (74). Many sinologists affirmed Confucianism's status as a philosophy to justify their perception of China as devoid of religious morality.[26] Morrison diverges from this logic. Despite Confucianism's failure as a religion, Morrison's appreciation of, even praise for, the ancient philosophy's moral teachings persists. In particular, he acknowledges that many Confucian values mirror Christian principles. For example, he describes how Confucius taught that "equals should do to others as they would others do to them, and always judge of other people by themselves" (71). When Mary aptly notes that this lesson echoes the Golden Rule, Morrison concurs that "some people in England mistake, when they say the 'golden rule' . . . is found no where but in the Testament" (72). By subtly acknowledging doctrinal similarities, Morrison lays the foundation for simultaneous critiques of both belief systems. In his discussion of war's virtues (or rather the lack thereof), Morrison describes the soldier's desire to acquire honor, even if that honor necessitates disobedience to God. Without any sense of a greater spiritual being, the Chinese are more willing, asserts Morrison, to sacrifice their lives and those of others. He argues that, "To be praised and admired by posterity, which is the only reward of virtue that Confucius acknowledged, is expected by the people who expose their lives, and try hard to kill others" (49). While these values may be condoned by Confucius, it does not follow that Christians are free of these faults. Rather Morrison perceives that "English heroes [similarly] look to Westminster Abbey as the reward of their victories" (49). Confucianism's flaws and virtues provide Morrison with a conduit through which to appraise British practices. As he advises his son, "If we look only at what is good in ourselves, and at what is bad in others, we shall, instead of *reforming*, feed our own pride, and be unjust to others. Confucius knew better than that" (102). In short, Morrison advocates that Anglo-Sino intellectual exchange could be mutually beneficial.

While Morrison identifies several Chinese cultural practices as exemplary and worthy of imitation, he is particularly fascinated with the Chinese appreciation of literature. Early in his dialogues, Morrison recognizes that European printing is indebted to the Chinese-developed technologies conveyed to Europe by Marco Polo in the late thirteenth century.[27] But more important to Morrison than printing technology is the way in which the Chinese government is grounded in literary learning. All government officials were required to pass imperial examinations that tested their knowledge of the aforementioned classical literary works. Morrison's exploration of Confucianism includes a description of the four classic Chinese texts, the *Si Shu*. He describes how, "[c]hildren at school commit to memory every word in these books: the master explains the meaning to them every day; and gentlemen who study to be made magistrates, write essays, or themes, taken from the Four Books. They contain perhaps nearly as much reading as a New Testament" (83–84). Morrison's son quickly observes that "the Chinese seem to study the Four Books of Confucius more than Christians do the Bible," an observation that Morrison confirms (84). This dedicated learning puts to shame general British religious practices. As Morrison radically states, "some are Christians only *in name*, not *in reality*" (82). Despite his dismissal of Confucianism as a religion, he appreciates the ways in which Confucianism is embedded in the very fabric of Chinese culture. Unlike the "nominal Christians in England," whom Morrison derides, the

Chinese fully embrace Confucian ideals through their dedicated learning and practice of the doctrine conveyed in the *Si Shu* (82).

Moreover, Morrison praises this classical education as one which eschews popular forms of literature, seemingly capitalizing upon contemporaneous British debates surrounding the novel.[28] He describes how Chinese families prevent their children from reading works, such as novels, which "may be misemployed, or perverted to mischievous purposes" (113). Morrison applauds the Chinese for recognizing that "it is far better for young persons to attend to the real business of life, and to substantial studies, than to the visionary dreams, and false representations of novelists" (113–114). Through Chinese customs, Morrison engages in British debates concerning high and low forms of literature and uses the Chinese example to advocate the publication and consumption of less frivolous literature. Morrison's representation of China as a highly literate and literary civilization engages with early nineteenth-century notions, largely attributable to Georg W. F. Hegel, that civilized nations were marked by substantial bodies of literature that delineated and defined the culture's historical development.[29] Accordingly, Morrison's descriptions establish China as a civilized nation whose customs are worthy of sustained study and emulation.

While I would argue that many Romantic-era sinologists strove to produce more objective representations of China grounded in empirical observations, Morrison's *China; A Dialogue* is radical precisely because he goes one step further by advocating the transference of cultural practices from China to Britain. Morrison's success in this endeavor relies on two characteristics of the physical book. First, both title page and preface make clear that this is an instructional text for British youth. Yet, Morrison is not training children to be future East India Company merchants or even missionaries but rather teaching them "to view all mankind as one family; and the whole world as one great domain or estate, in the different parts of which our Heavenly Father has placed his children" ("Preface"). To this end, Morrison assumes a justifiably neutral attitude toward the Chinese. Second, the book was published anonymously by an

"Anglo-Chinese." Morrison did not avoid acknowledging the work as his own; the copy held at Indiana University's Lilly Library contains his personal inscription. Instead, in assuming the moniker "Anglo-Chinese," Morrison reflects the degree to which his scholarship and his identity are products of a global contact zone. He essentially rejects a single national identity. In fact, publishing the work anonymously provides him with the license to applaud Chinese culture, an attitude not readily accepted by the London Missionary Society, the East India Company, or the domestic British reader. Anonymity allows Morrison to develop a multifaceted representation of China that fails to conform to and even challenges emergent nineteenth-century British imperial ideology.

Notes

1. See Kingsley Bolton, Introduction to *A Vocabulary of the Canton Dialect*, by Robert Morrison (1828; repr. London: Ganesha; Tokyo: Edition Synapse, 2001): v–xliv, v–vi; Ka Lun Leung, "Missions, Cultural Imperialism, and the Development of the Chinese Church," in *After Imperialism: Christian Identity in China and the Global Evangelical Movement*, ed. Richard R. Cook and David W. Pao (Eugene, OR: Pickwick, 2011), 24–26.

2. James L. Hevia, *English Lessons: The Pedagogy of Imperialism in Nineteenth-Century China* (Durham, NC: Duke University Press; Hong Kong: Hong Kong University Press, 2003), 20.

3. Peter J. Kitson, *Forging Romantic China: Sino-British Cultural Exchange, 1760–1840* (Cambridge: Cambridge University Press, 2013), 73.

4. Suzanne W. Barnett, "Silent Evangelism: Presbyterians and the Mission Press, 1807–1860," *Journal of Presbyterian History (1962–1985)* 49, 4 (1971), 288–289.

5. Bolton, Introduction, vi.

6. According to the *Oxford Dictionary of National Biography*, Morrison's legacy is that of "a scholarly facilitator of cultural exchange between Europe and China" (R. K. Douglas, "Morrison, Robert (1782–1834)," rev. Robert Bickers, *Oxford Dictionary of National Biography* (Oxford: Oxford University Press, 2004) online edition, May 2007; https://www

.oxforddnb.com/view/10.1093/ref:odnb/9780198614128.001.0001/odnb
-9780198614128-e-19330).

7. George Thomas Staunton, quoted in Eliza Morrison, *Memoirs of the Life and Labours of Robert Morrison* (London, 1839), 2:442.

8. Christopher A. Daily, *Robert Morrison and the Protestant Plan for China* (Hong Kong: Hong Kong University Press, 2013), 7.

9. Barnett, "Silent Evangelism," 294.

10. Shu-mei Shih, "The Concept of the Sinophone," *PMLA* 126, no. 3 (2011), 711.

11. Bolton, Introduction, xvi.

12. Robert Morrison, *Dialogues and Detached Sentences in the Chinese Language; With a Free and Verbal Translation in English. Collected from Various Sources. Designed as an Initiatory Work for the Use of Students of Chinese* (Macao, 1816); and Robert Morrison, *China; A Dialogue, For the Use of Schools: Being Ten Conversations between a Father and His Two Children, Concerning the History and Present State of That Country* (London, 1824). Subsequent references to *China; A Dialogue* will give page numbers in parentheses.

13. Mary Louise Pratt, *Imperial Eyes: Travel Writing and Transculturation*, 2nd ed. (London: Routledge, 2008), 74.

14. Pratt, *Imperial Eyes*, 82.

15. Pratt, *Imperial Eyes*, 9.

16. George Thomas Staunton, trans., *Ta Tsing Leu Lee: Being the Fundamental Laws, and a Selection from the Supplementary Statutes of the Penal Code* (1810) (Cambridge: Cambridge University Press, 2012), 532–537.

17. Chris Murray, "'Wonderful Nonsense': Confucianism in the British Romantic Period," *Interdisciplinary Literary Studies* 17, no. 4 (2015), 606.

18. Murray, "Wonderful Nonsense," 602.

19. Morrison, *Dialogues and Detached Sentences*, v–vi.

20. Morrison, *Dialogues and Detached Sentences*, 2.

21. Morrison, *Dialogues and Detached Sentences*, 4.

22. See Robert Markley, *The Far East and the English Imagination, 1600–1730* (Cambridge: Cambridge University Press, 2006), 1–29. In the early eighteenth century, Markley argues, "the Middle Kingdom symbolized the very principles of sociopolitical stability and transcultural value on which European elites depended" (3).

23. Morrison, *Dialogues and Detached Sentences*, 64–67.

24. The Enlightenment notion of stadial theory, often termed "conjectural history," presumed that all societies experience the same stages of civil development, though at different historical times. According to Adam Smith, "There are four distinct states which mankind pass thro:—1st, the Age of Hunters; 2dly, the Age of Shepherds; 3dly, the Age of Agriculture; and 4thly, the Age of Commerce" (*The Glasgow Edition of the Works and Correspondence of Adam Smith*, ed. R. L. Meek, D. D. Raphael, and P. G. Stein, vol. 5, *Lectures on Jurisprudence* [Oxford: Clarendon, 1978], 14). Whereas nineteenth-century Britain was understood to have reached "the Age of Commerce," China, as perceived by nineteenth-century Westerners, remained in "the Age of Agriculture." For further scholarly discussion of stadial theory's colonial impact, see Larry Wolff and Marco Cipollini, eds., *The Anthropology of the Enlightenment* (Stanford, CA: Stanford University Press, 2007).

25. John Barrow, *Travels in China: Containing Descriptions, Observations and Comparisons, Made and Collected in the Course of a Short Residence at the Imperial Palace of Yuen-Min-Yuen, and on the Subsequent Journey through the Country from Pekin to Canton* (1804) (Cambridge: Cambridge University Press, 2010), 73.

26. See William B. Langdon, *Ten Thousand Chinese Things. A Descriptive Catalogue of the Chinese Collection, Now Exhibiting at St. George's Place, Hyde Park Corner, London, with Condensed Accounts of the Genius, Government, History, Literature, Agriculture, Arts, Trade, Manners, Customs, and Social Life of the People of the Celestial Empire*, 12th English ed. (London, 1842), vii.

27. For more on the iconic role of Marco Polo in British sinology, see Jennifer L. Hargrave, "*Marco Polo* and the Emergence of British Sinology," *SEL: Studies in English Literature, 1500–1900* 56, no. 3 (2016), 515–537.

28. For examples of contemporaneous debates surrounding the novel, see Charlotte Lennox, *Entertaining History of the Female Quixote, or The Adventures of Arabella* (London, 1752); or Henry Mackenzie, *The Lounger* 20 (June 18, 1785), 168–175.

29. Georg W. F. Hegel, *Lectures on the Philosophy of World History. Introduction: Reason in History*, ed. Johannes Hoffmeister, trans. H. B. Nisbet (Cambridge: Cambridge University Press, 1975), 12.

6

At Home with Empire?

Charles Lamb, the East India Company, and "The South Sea House"

JAMES WATT

Throughout his correspondence and the essays that he wrote in the persona of Elia, Charles Lamb reflected—by turns directly and obliquely—on his experience of working at the Leadenhall Street headquarters of the East India Company, where from 1792 to 1825 he was a clerk in the accountant's office. In some respects Lamb might be regarded as a Cockney version of the renowned orientalist Sir William Jones, who sailed for India in 1784 to take up a position as a judge in the Supreme Court of Bengal, and who over the next decade produced in his spare time numerous important works of scholarship, literary translation, and original poetic composition. Both men were employees of the Company, in London and Calcutta respectively, but both would become best known for the work that they did beyond their immediate service of the Company's interests. Lamb, like Jones, additionally seems to have conceived of a "second self" that was not accountable to authority or responsible to other people, and he imagined literary composition more or less as Jones saw scholarly research, as (in the latter's words) a "relief" from "severer employment in the discharge of publick duty."[1] If Lamb in his own distinctive and haphazard way can sometimes appear to be almost as learned a figure as the polymathic Jones, however, there are significant differences between the two men as well. Leaving aside the obvious matter of geographical location, the most salient of these for the purposes of this essay is that whereas Jones carried his sense of responsibility to the Company into the other diverse forms of writing that he did, Lamb chafed against his relatively lowly role as a bookkeeper and shifted between protesting at the discipline and routine of office life and attempting to conceive of an imaginative space free from its rigors.

Looking at Lamb alongside Jones brings into sharper focus the character of his position as a metropolitan author whose professional work—a much more mundane form of "writing" than the kind he aspired to produce—implicated him in what H. V. Bowen has termed "the business of empire."[2] Lamb's position is further illuminated when it is compared with that of another London-based Company man, the higher-ranking Thomas Love Peacock, who referred to the post in the examiner's office at India House that he acquired in 1819 as offering "an employment of a very interesting and intellectual kind, connected with finance and legislation, in which it is possible to be of great service, not only to the Company, but to the millions under her dominion."[3] Whereas Peacock in this quotation presents himself as integrated into a network of empire, gaining fulfillment from an administrative role that harnessed his capacities, Lamb might be regarded as someone who was "networked," by virtue of his occupation, but who nonetheless failed or refused to identify any larger purpose to make his daytime labors seem meaningful. As I will argue in this essay, Lamb in his correspondence often paraded an anticosmopolitanism that could sometimes take the form of a wholesale repudiation of the system of imperial commerce that he served. In the first of his Elia essays, "Recollections of the South Sea House" (published in the *London Magazine* in August 1820), however, as I will go on to show, Lamb also explored the experience of work as service in a more complex and wide-ranging way than this. Though Elia's elegiac remembrance of his former colleagues may be read as a product of Lamb's escapist yearning, I suggest that the essay's representation of their eccentricity and its setting up of a distinction between South Sea House in the past and India House in the present additionally help to stage a new metropolitan sense of a seemingly unproblematic arm's-length relationship between Britain and its expanding Eastern empire.

Lamb often complained about his inability to find any respite from the demands of his work at India House, and in a letter that he wrote to Wordsworth in September 1814 he claimed that his duties consumed even his dream-life: "my sleep is nothing but a succession of dreams of business I cannot do."[4] In a letter to the poet Matilda Betham in 1815, he protested in more specific terms about how his creative ambition was compromised by the mechanical ledger-keeping routines that occupied his time: "[w]hy the devil am I never to have a chance of scribbling my own free thoughts, verse or prose, again? Why must I write of Tea & Drugs & Price Goods & bales of Indigo[?]"[5] Lamb's "disidentification" with the Company was such that he sometimes referred to his employment almost as if he saw himself as a colonial subject rather than, like Jones, someone who was able to associate personal fulfillment and Company gain, or "delight and advantage."[6] He periodically styled himself as captive to his employer, functioning in a state of "official confinement": "[t]hirty years have I served the Philistines," he wrote to Wordsworth in March 1822, "and my neck is not subdued to the yoke."[7] He more explicitly presented his Leadenhall Street labors as a kind of thralldom in a letter to "Quaker poet" Bernard Barton in September 1822, in which, speaking as if he had been seized by Barbary pirates, he stated that "I am, like you, a prisoner to the desk. I have been chained to that galley thirty years."[8]

This allusion to Barbary piracy is idiosyncratic not least because of its archaism, but it nicely captures the way in which Lamb—often, as here, rather stagily—imagined the hostility beyond his immediate milieu of pleasingly familiar reference points and associations. David Higgins argues that Lamb constructed an "obsessively localised self that is uneasy to the point of morbidity in its apprehension of the exotic," and he cites a letter that Lamb wrote to Robert Southey in May 1815 in which he claimed that the crusading rhetoric of his friend's epic poem *Roderick, the Last of the Goths* (1814) allowed him as a reader to remain within "the pale of . . . old sympathies."[9] While characteristically arch, Lamb's acknowledgment of the narrowness of his horizons here appears to be defensive too, since in invoking "the pale" he alludes to the embattled English settlements in Ireland that were first established during the reign of Henry II. Whereas Jones declared his desire to "know" the East better than anyone had done before, Lamb, in his letters and his Elia essays, frequently performed a lack of awareness of or interest in the wider world.[10] In another letter of 1815, written on Christmas Day, Lamb told his friend Thomas Manning, a Company employee and sinologist based in Canton, that while so much had happened in "the western world" in recent years, he had been wasting his time "settling whether Ho-hing-tong should be spelt with a—or a—."[11] What is especially striking about this is that Lamb expressed such a derisive sense of Chinese oddity almost a decade after Manning first sailed for Canton, and despite their regular correspondence during that period.[12] For all that he was in a position to know better, then, Lamb maintained a willful ignorance where China was concerned. Throughout his writing, indeed, he often assumed comparably uncosmopolitan postures, even as his bookkeeping work in its own small way helped to facilitate Britain's East India trade and its commercial intercourse with Asia more generally.

Higgins emphasizes that Lamb's complaints about work and his anxious relation to anything beyond his "old sympathies" sometimes overlapped with or shaded into a more politicized perspective on commercial modernity.[13] In a letter that he wrote to Wordsworth in April 1815, for example, Lamb lamented his inability to be as productive as his friend and then explained why this was the case:

these "merchants and their spicy drugs" which are so harmonious to sing of, they lime-twig up my poor soul and body, till I
should forget I ever thought myself a bit of a genius! I can't even
put a few thoughts on paper for a newspaper. I "engross," when I
should pen paragraph. Confusion blast all mercantile transactions, all traffick, exchange of commodities, intercourse between
nations, all the consequent civilization and wealth and amity
and link of society, and getting rid of prejudices, and knowledge
of the face of the globe—and rot the very firs of the forest that
look so romantic alive, and die into desks.[14]

As in the letter to Matilda Betham cited above, Lamb in this passage claims that the need to perform one kind of apparently labo-
rious and mind-numbing writing ("engrossing") stands in the way
of him producing another of a more congenial variety; depicting
himself as literally stuck, "soul and body," Lamb again refers to
deskwork as a state of captivity.[15] After thus registering another
complaint about his employer thwarting his aspirations, however,
Lamb also moves beyond the issue of his own lack of suitability
for office life.[16] By citing Milton's description of Satan's flight from
Hell in book 2 of *Paradise Lost* ("merchants and their spicy drugs"
is a misquotation of a passage alluding to the Dutch spice trade),
he presents commercial modernity as a fall from humankind's
natural state—a descent figured by the contrast between the live
"firs of the forest" and the "dead" desks of clerks such as himself
that are produced from them.[17] While he eschews "knowledge of
the face of the globe" in terms familiar from his other writings,
Lamb at the same time denounces the whole system of exchange
with which any such knowledge would have been intimately connected, with the effect of suggesting that an obverse ignorance
might be a principled position. In the process he draws attention
to the gulf between the ideology of "le doux commerce" and his
own experience as a functionary of the Company, rehearsing with
ironic detachment the various expressions of the argument that the
agency of commerce united the peoples of the world in peaceful
and mutually beneficial interrelationship.

Lamb invoked the idea of global trade as Fall on a number of occasions, as in a letter of September 1805 to William and Dorothy Wordsworth in which he represented the "pretence of Commerce allying distant shores, promoting and diffusing knowledge, good &c.—" as a Satanic ruse, "the invention of the Old Teazer who persuaded Adam's Master to give him an apron and set him houghing."[18] He returned to this mock-Miltonic register in his account of the "Beasts" ("Accountants, Deputy Accountants") to whom he was responsible at India House, telling Mary Wordsworth in February 1818 that while the "dear abstract notion of the East India Company, as long as she is unseen, is pretty, rather Poetical," he "loathe[d] and detest[ed]" the actual manifestation of her presence as much as "the Scarlet what-do-you-call-her of Babylon."[19] This complaint against the Company is expressed in characteristically hyperbolic terms, and the immediate occasion for it appears to have been outrage at the revocation of a customary right to go home early on a Saturday rather than any ethical objection to the effects of empire, whether in Britain or overseas; once again Lamb here is agitated by the discipline imposed on him at India House.[20] The idea of "Babylon" invoked by Lamb has a suggestive resonance, however, not just in connection with Milton's orientalized Pandemonium, but by way of analogy for the condition of modern Britain too.[21] As Gregory Dart has noted, parallels between Britain and Babylon (made explicit in the apocalyptic scenes painted by John Martin) were to become especially frequent around the time of the death of George III in 1820.[22] Although Lamb often rehearsed "end of empire" discourse in a jocular manner (in the Christmas 1815 letter to Thomas Manning cited above, for example, he told his Canton-based friend that "St. Paul's church is in a heap of ruins"), it is a recurrent feature of his writing nonetheless.[23]

In much of Lamb's writing, then, a desire to be free from the demands and determinations of work appears to be bound up with a larger sense of (albeit implicit) resistance to the commercial empire that he served; certainly, Lamb's letters often take a skeptical view of any grand narrative of Company-sponsored "improvement" of the sort articulated by Thomas Love Peacock or by another famous India House employee, James Mill. In the rest of this essay, however, I want to consider how Lamb might all the same be regarded as a kind of "imperial man," at least insofar as, as H. V. Bowen has argued, he was—despite his protestation to the contrary—among those whose "minds and skills were being applied to the government of a vast territorial empire as well as to the continuing management of an expanding trade."[24] Lamb's depiction of the South Sea House in his essay of that title is especially interesting here because it helps us to think about both the idea of the invisible— or "unseen"—workings of the East India Company (which Lamb alluded to in his February 1818 letter to Mrs. Wordsworth) and the potential for London's position as a metropolis that was partly at least "made" by empire to remain similarly unexamined. For all that Lamb himself sometimes railed against global commerce and signaled his wariness about anything beyond the pale of his "old sympathies," "The South Sea House," I will claim, suggestively reflects on the subjectivity of Britons who may have been able to be "at home with empire" (in the words of Catherine Hall and Sonya O. Rose), by "simply assuming it was there, part of the given world that had made them who they were."[25]

The South Sea House was the headquarters of the South Sea Company, which was popularly associated with the speculative "bubble" of 1720, itself taken by many to symbolize the inherent instability of the new world of the eighteenth-century credit economy; the South Sea Company also held Britain's monopoly right to sell slaves to Spanish America, as granted by the Asiento contract of 1713. In the essay titled "The South Sea House," Lamb's narrator Elia describes this now vacant "house of trade" as a "magnificent relic!" which provides at once an oasis of calm and a vehicle for pleasurable contemplation.[26] He presents the South Sea House in terms that chime with Lamb's own expressions of a frustrated desire for freedom from the necessity of paid employment: "to such as me, old house!," he writes, drawing attention to the peculiarity of his own sensibility, "there is a charm in thy quiet: a cessation—a coolness from business—an indolence almost cloistral—which is delightful!" (2). Elia adopts the language of Mr. Spectator at the Royal Exchange (in *Spectator* no. 69), only to locate the existence of the South Sea House as "a centre of busy interests" firmly in the past: "[t]he throng of merchants *was* here."[27] Though he acknowledges that "some forms of business are still kept up" (because the South Sea Company continued to manage part of the national debt), he states that the "soul" of the South Sea House is "long since fled" (1) nonetheless. The living death of the South Sea House and its place as a "memorial . . . in the very heart of stirring and living commerce" make it available to be read, as Higgins argues, as a reminder of the "obsolescence and destruction that awaits the grandest imperial projects"—thereby implicating India House and also London as a whole, at a time when numerous schemes were being proposed to create a more impressive cityscape appropriate to Britain's status as the world's preeminent power.[28] In line with such a reading, Elia's account of his solitary pacing of the "great bare rooms and courts" (2) of the dust-layered South Sea House might be seen to allude to a near contemporary "ruins of empire" text such as Anna Letitia Barbauld's "Eighteen Hundred and Eleven" (1812), in which the narrator, a "last man," traverses a Britain now vacated by the "Genius" that had once presided over it.[29]

This reading of the South Sea House as a generally applicable memento mori is especially plausible in view of the critically detached position on the East India Company that Lamb so often adopted in his correspondence, as discussed above. If Lamb chose to commemorate the centenary of the South Sea Bubble, as Gerald Monsman suggests, then he may indeed have wanted to indicate the contemporary relevance of this history.[30] Elia's subsequent recollection of the "[o]dd fishes" with whom he worked "forty years back" (3) offers an alternative juxtaposition of past and present, however, introducing the reader to a group loosely modeled on the Spectator Club, as Monsman points out, and perhaps also referencing the gaggle of clerks depicted in the foreground of Augustus Charles Pugin and Thomas Rowlandson's "The South Sea House," one of the plates in Rudolph Ackermann's *The Microcosm of London* (1808–1810).[31] Elia states that his former colleagues "had an air very different from those in the public offices that I have had to do with since" (3) (thereby reminding us of Lamb's complaints about the workplace discipline that he experienced at India House), and he additionally describes them as men who—unlike Lamb by his own account—managed to leave the world of business behind them at the end of the working day. The first of the shades he summons is that of "one Evans, a Cambro-Briton" (3), whom he remembers as becoming suddenly enlivened when his daytime labors ceased: "[t]hen was his *forte*, his glorified hour! How would he chirp, and expend, over a muffin!" (4).[32] Elia then invokes a more nonchalant figure, John Tipp, who (unlike Evans's deputy, Thomas Tame, who "had the air and stoop of a nobleman" [4]) "neither pretended to high blood, nor in good truth cared one fig about the matter." Recalling that Tipp's out-of-hours "hobby" was music, Elia presents him as playing host to a fortnightly concert of kindred spirits: "'sweet breasts' . . . culled from club-rooms and orchestras—chorus singers—first and second violoncellos—double basses—and clarionets—who ate his cold mutton, and drank his punch, and praised his ear" (5).

Elia states that when Tipp played his fiddle he "did . . . scream and scrape most abominably" (5), and for Monsman there is a pathos in the way in which Elia portrays Tipp and others as "victims of their illusions" even as he draws attention to the unreliability of his own reveries too; "[r]eader, what if I have been playing with thee all the while" (8), he cautions toward the close.[33] The illusions of Tipp (whose "fine suite of official rooms," we are told, "were enough to enlarge a man's notions of himself that lived in them" [5]) may further however be understood in relation to William Hazlitt's account, in "On Londoners and Country People" (1825), of the "positive illusions"—not simply reducible to false consciousness—of the proud and self-possessed Cockney.[34] In view of Lamb's extensive reflection on the experience of work, in his correspondence and his essays, it is also striking that Elia describes Tipp as protective of his leisure hours but "quite another sort of creature" in the South Sea House, where he was the embodiment of a notion of professional service avowedly divorced from the world outside of the office. In Tipp's company, Elia states, "all ideas, that were purely ornamental, were banished. You could not speak of any thing romantic without rebuke. Politics were excluded. A newspaper was thought too refined and abstracted. The whole duty of man consisted in writing off dividend warrants." Elia presents Tipp as one who regarded "'an accountant the greatest character in the world, and himself the greatest accountant in it'" (5), and his claim that, for Tipp, "[t]he fractional farthing" was "as dear . . . as the thousands which stand before it" nicely conveys the latter's understanding of figures on balance-sheets as referring only to themselves. He goes on to add that Tipp never "for lucre, or for intimidation . . . forsook friend or principle" (6), but this devotion to principle is defined as one that is restricted to the technicalities of accounting practice rather than concerning itself with any wider public sphere.

It might be argued that Tipp is depicted as functioning in a kind of asocial vacuum in this way because he is associated with the near-defunct South Sea Company, and therefore in effect could only ever have played at, or gone through the motions of, being an accountant: Tipp and his fellow clerks "had not much to do" (3), Elia states. By virtue of his ethic of professional service and his clear demarcation of work and leisure time, however, Tipp intriguingly anticipates the much better-known figure of Mr. Wemmick in Dickens's *Great Expectations* (1861). Dickens's portrait of Wemmick satirizes his "Cockney false consciousness," Gregory Dart argues, yet also registers the appeal of the "harmless fantasy world" that he creates for himself and his father; it additionally invites us to recognize his ability imaginatively to detach his suburban idyll from the source of the income that sustains it, his work as a clerk to the criminal lawyer Jaggers (who represents, among others, the convict Abel Magwitch).[35] Albeit that Tipp is not a suburban Cockney of the same stamp as Wemmick, reading Tipp alongside Wemmick brings into focus Elia's description of him as upholding an ideal of "commercial" honor—as in his careful treatment of "the fractional farthing" (6)—yet also separating such duty from any consideration of the realities of the commercial exchange to which numbers on balance sheets correspond. Elia presents "[t]he striking of the annual balance in the company's books" as occupying Tipp's "days and nights for a month previous" (5), even though he yearned for busier times, and "a return of the old stirring days when South Sea hopes were young!" (6). While his sense of "the old stirring days" identifies Tipp as himself a nostalgist, like Elia, his apparent allusion to the pre-bubble period ("when South Sea hopes were young!") also refers to a time when trading in slaves was one of the South Sea Company's key areas of activity. The South Sea Company administered Britain's monopoly right to sell slaves to Spanish America, as noted above, but this is a history that the memories of Tipp and Elia significantly neglect to acknowledge.

Elia's recollection of another South Sea House clerk, Henry Man, provides a further demonstration of how certain forms of nostalgia for the past may help to distance potentially troubling aspects of its reality from current consideration. He presents Man as, in contrast to Tipp, someone who was eager to learn about matters outside of the workplace: "great thou used to be in Public Ledgers, and in Chronicles" (7), he states. In line with his claim to be speaking of a time "forty years back" (3), Elia locates Man historically by referring to him as an expert "upon Chatham, and Shelburne, and Rockingham, and Howe, and Burgoyne, and Clinton, and the war which ended in the tearing from Great Britain her rebellious colonies,—and Keppel, and Wilkes, and Sawbridge, and Bull, and Dunning, and Pratt, and Richmond,—and such small politics" (7). The way in which Elia breaks off the list of names in this quotation (after "Clinton") only then to resume it (with "Keppel") captures the zealousness of Man's fixation upon the politics of the period: once he started talking it was difficult to get him to stop, it is suggested. By thus perhaps portraying him as a "victim"—to adopt Monsman's term—of his unfocused and susceptible curiosity, this account of Man appears to be in the tradition of "newsmonger" satire, belittling the inappropriate interest in current affairs of those with ideas above their station.[36]

Through its depiction of Man, however, Lamb's essay can once more be seen indirectly to consider the nature of the imperial metropole's contemporary relation to the wider world. While some of the names in the list above call up an earlier period in the history of reformist politics (and perhaps signal Lamb's—nostalgic—sense that this was a more progressive moment), the reference to Britain's war with its American colonies provides a reminder of a pivotal episode in the nation's recent history of empire. Karen Fang has claimed that Elia's "aesthetic contemplation" of the South Sea House "evacuates [the] formerly controversial history" of Britain's involvement in the slave trade (once conducted via the South Sea Company), and the double meaning of "small politics" in the quotation cited is suggestive of a comparable process of "evacuation" having already taken place.[37] On the one hand the phrase "small politics" functions as an ironic recognition of the trauma of Britain's war with its American colonies and loss of its "first" empire, which for some at the time appeared to presage the nation's demise as an imperial power.[38] On the other, though, it offers a fitting sense of how, taking a longer view, defeat in America did indeed come to seem a "small" and ultimately forgettable matter as Britain subsequently extended its sovereignty and influence, and developed its "second" empire, in the East. The idea of "small politics" moreover attests to the way in which the concept of empire itself became steadily less controversial in metropolitan circles after the intense introspection that initially accompanied defeat in America. As Nicholas B. Dirks has argued, the long-drawn-out trial of the former governor general of Bengal Warren Hastings (which concluded with his acquittal in 1795) had the effect of purging the "scandal" of Indian empire.[39] The subject of British India is more or less absent from Lamb's writing, and it is fair to say that the realities of Company rule in the early nineteenth century—whether the Anglo-Maratha wars, further annexations of territory, or the establishment of relations of suzerainty with princely states—generated little commentary in the national culture more broadly.[40] Absent too from Lamb's writing—though an increasingly important aspect of the "unseen" business of the Company at this time—is the illegal trade by which its merchants exchanged Indian opium for Chinese tea, and which Lamb himself probably played a part in recording in his work at India House.[41]

Lamb's essay to some extent remains embedded in the civic humanist rhetoric of decline and fall, as indicated above, and in its account of the South Sea House as a "magnificent relic!" (2) it may be read as presenting (in the words of Higgins again) "a London that is fundamentally shaped by imperial power and trade, but that is also haunted by the prospect of its own decay."[42] Keeping in mind both the connection of the South Sea House to the slave trade and the depiction of Henry Man's obsession with the politics of the American war, however, it may further be argued that the essay bears witness to a transition from one phase of empire to another—where the former was in its own time contentious but the latter is seemingly uncontroversial, not least since it is imaginatively remote from, even as it is bound up with, everyday life in the imperial metropolis.[43] Elia can indeed be seen to reflect on the condition of being "at home with empire," by initially drawing attention to his audience's sense of their familiar experience and routine. He begins by addressing a "reader" whose daily "passage from the Bank [of England]" takes him past the South Sea House, and he goes on to describe this one-time "house of trade" (1) as situated "in the very heart of stirring and living commerce,—amid the fret and fever of speculation—with the Bank, and the 'Change, and the India-house about [it], in the hey-day of present prosperity" (2). As much as the South Sea House itself is portrayed in terms of a "ruins of empire" discourse, then, the essay here returns to the urtext of eighteenth-century commercial empire, Joseph Addison's *Spectator* no. 69, to record the economic activity which flourishes around it. In the process it alludes to the prospectus of the periodical in which it was published, the *London Magazine*, which announced as one of its aims that of "convey[ing] the very 'image, form, and pressure' of that *'mighty heart'* whose vast pulsations circulate life, strength, and spirit, throughout this great Empire."[44] If Elia initially states of the South Sea House that "the throng of merchants was here" and that both its "quick pulse of gain" and its "soul" (1) have long since fled, he goes on to acknowledge that Londoners are habituated to a greater dynamism than this in the present.

This is not to argue that a narrative of "present prosperity" simply overwrites a narrative of decay in Lamb's essay: the "fever of speculation" is an obtrusive phrase which specifically nods to the bubble of 1720 (thereby collapsing the gap between past and present) and more generally, with its connotations of disease and illness, cautions that the circulatory system of "stirring and living commerce" (2) may be subject to disruption if not breakdown; Lamb sometimes thought of "this mercantile city" as an essentially alien environment, and Elia clearly prefers quiet to bustle.[45] I do want to suggest however that the essay might at least be seen to offer a provisional negotiation of such "anxieties of empire," by forestalling any idea that India House would inevitably go the way of its *poor neighbour out of business*" (2), the South Sea House. In this context it is useful to consider an earlier poem by Anna Letitia Barbauld, her 1791 "Epistle to Wilberforce," which warns that while pleasure-seeking Britons remain oblivious to their nation's impending fall, an irreversible process of corruption, figured in part as a return of the repressed, is already well underway: "By foreign wealth are British morals chang'd, / And Afric's sons, and India's, smile aveng'd."[46] With these lines from Barbauld's poem in view, it seems fair to say that Lamb's essay contains any such civic humanist narrative of decline and fall, in part because the now vacant South Sea House provides a pleasing rather than troubling spectacle, and in part because this former commercial headquarters is "[s]ituated" (2) in an imperial metropolis which is at least as energetic as that of the London of *Spectator* no. 69 but which at the same time bears no comparable trace of the other; while Elia briefly describes more or less the same mercantile milieu as Mr. Spectator—the area around "the Bank, and the 'Change"—he notably eschews the cosmopolitanism of Addison's persona, a self-styled "Citizen of the World" who celebrates the "rich . . . Assembly of Country-men and Foreigners" that congregate at the Royal Exchange to do business together.[47]

As recent scholarship inspired by the "new imperial history" has shown, the London of Lamb's time would unquestionably have borne the imprint of being (to recall the *London Magazine*'s Wordsworth-derived phrase) the "mighty heart" of an expanding empire. To consider the population of the metropolis, one of the most visible markers of London's connection to the wider world, high-profile visitors in the decade or so prior to Lamb's composition of his Elia essays included the Indian scholar Mirza Abu Taleb Khan, author of a widely read account of his *Travels* (1810), and the Persian ambassador Mirza Abul Hassan Khan, to whom Lamb referred in a January 1810 letter to Thomas Manning as "the principal thing talked of now;" "[t]he people call him Shaw Nonsense," he noted without any obvious disapproval.[48] William Hazlitt's essay "The Indian Jugglers" (1821–1822) is premised on the idea that the spectacle it describes would have been an unremarkable experience for Londoners: the jugglers of the title provide Hazlitt with the point of departure for a meditation on the relationship between "intellectual" and "mechanical" forms of human achievement.[49] Dean Mahomet opened his "Hindostanee Coffee House," Britain's first Indian restaurant, in 1810 to cater for colonial returnees eager to carry on eating the same cuisine that they had enjoyed in India, and from the 1820s onward, as Daniel E. White has shown, the "Little Bengal" community of Anglo-Indians was firmly established in Marylebone and Mayfair, as "a unique space of exile in the very heart of home."[50]

Where the world of "Little Bengal" is concerned, "The South Sea House" refers to its nominal addressee's north-south movement between Dalston and the City rather than an east-west movement between the City and Marylebone and Mayfair, and it may be that this Indianized community was socially as well as geographically separated from Lamb's own everyday life and familiar routines. If we can only speculate now about Lamb's actual experience of an increasingly cosmopolitan metropolis, however, we can say that across his writing the India House employee Lamb did little directly to engage with the reality of how London was, as Jon Klancher puts it, "emerging imperially as [a] world city."[51] One way of explaining this absence is to acknowledge that Lamb in his correspondence often adopted positions that were critical of commercial modernity (notably in relation to his experience of work), and that in his essays he frequently used his Elia persona to dwell upon the seductive pull of the past.[52] It is also the case, though, as I have suggested, that by representing an accountant for whom "form was every thing" (6) and by locating the scandal of empire "forty years back" (3), as well as by addressing a "reader" whose sense of routine appears to make him oblivious to his surroundings, "The South Sea House" reflects on how its audience—and Lamb himself—may have been able to conceive of the "present prosperity" (2) in part symbolized by India House as independent of the network of empire that made it possible. I will conclude by returning to *Great Expectations*, a novel which famously uses its account of Pip's connection to Magwitch in order to allegorize the relationship between Britons' sense of themselves and their colonial wealth. While Pip's inability to escape from the "taint of poison and crime" invites readers to consider submerged links between metropole and colony, the role that Pip acquires in the service of the shipping company Clarriker and Co. in Egypt, after the death of Magwitch, locates him squarely within a network of imperial commerce; unlike Mr. Wemmick, he is open about the nature of his position, stating that he and Herbert Pocket "were not in a grand way of business, but . . . had a good name, and worked for our profit, and did very well."[53] Albeit that it does briefly link "present prosperity" to the East India Company, "The South Sea House" by contrast keeps Britain's global commerce at an imaginative distance. In the process, as I have tried to argue in this essay, it stages an ideological formation which appears able to conjoin being "at home with empire" and not thinking very much about empire at all.

Notes

1. Sir William Jones, "To Samuel Davis [September 21, 1786]," in *Letters of Sir William Jones*, ed. Garland Cannon (Oxford: Clarendon, 1970), vol. 2, 705.

2. H. V. Bowen, *The Business of Empire: The East India Company and Imperial Britain, 1756–1833* (Cambridge: Cambridge University Press, 2006).

3. Cited in Bowen, *Business of Empire*, 146.

4. Charles Lamb, "To William Wordsworth [September 19, 1814]," in *The Works of Charles and Mary Lamb*, ed. E. V. Lucas (London: Methuen, 1905), 6:443.

5. Lamb., "To Matilda Betham [1815]," in *Works*, 6:478. Lamb prefaced this question by stating "God bless you (tho' he curse the India House & fire it to the ground) . . . [.]"

6. This is a recurrent phrase in Jones's writing—see for example his inaugural address to the Asiatic Society of Bengal, published as *A Discourse on the Institution of a Society for Enquiring into the History, Civil and Natural, the Antiquities, Arts, Sciences, and Literature of Asia*" (London: T. Payne, 1784), 10.

7. Lamb, "To William Wordsworth [March 20, 1822]," in *Works*, 7:563.

8. Lamb, "To Bernard Barton [September 11, 1822]," in *Works*, 7:572.

9. David Higgins, *Romantic Englishness: Local, National, and Global Selves, 1780–1850* (Basingstoke: Palgrave Macmillan, 2014), 130; Lamb, "To Robert Southey [May 6, 1815]," in *Works*, 6:465.

10. Jones stated that "it is my ambition to know *India* better than any other European ever knew it," "To the second Earl Spencer [August 17, 1787]," in *Letters*, 2:751.

11. Lamb, "To Thomas Manning [December 25, 1815]," in *Works*, 6:481.

12. In an earlier letter that he wrote to Manning, Lamb told him: "When you get to Canton, you will most likely see a young friend of mine, Inspector of Teas, named Ball. He is a very good fellow and I should like to have my name talked of in China." "To Thomas Manning [May 10, 1806]," in *Works*, 6:348; the Inspector of Teas post was created in order to monitor the quality of tea exported from China, and Lamb's reference to Ball's role suggests that he had some awareness of the way in which East India Company trade worked.

13. Higgins, *Romantic Englishness*, 140–144.

14. Lamb, "To William Wordsworth [April 28, 1815]," in *Works*, 6:463.

15. The *OED* defines "engrossing" as writing "in a peculiar character appropriate to legal documents" and the practice of "includ[ing] in a list," and both of these senses of the word are in play here.

"[L]ime-twig" is an ironic nod to Coleridge's poem "This Lime-Tree Bower My Prison" (written in 1797), which was initially subtitled "A Poem Addressed to Charles Lamb, of the India House, London," and which famously refers to Lamb as "[i]n the great City pent," Samuel Taylor Coleridge, *The Complete Poems*, ed. William Keach (Harmondsworth: Penguin, 1997), 139.

16. "I am not fit for an office," "To William Wordsworth [September 19, 1814]," in *Works*, 6:444.

17. Balachandra Rajan, *Under Western Eyes: India from Milton to Macaulay* (Durham, NC: Duke University Press, 1999), 53; Higgins, *Romantic Englishness*, 144–145.

18. Lamb, "To William Wordsworth [September 28, 1805]," in *Works*, 6:317.

19. Lamb, "To Mrs William Wordsworth [February 18, 1818]," in *Works*, 6:513.

20. In the letter cited above, Lamb added that "I thought, after abridging us of all our red letter days, they had done their worst, but I was deceived in the length to which Heads of offices, those true Liberty haters, can go"; as in a letter to Bernard Barton in January 1823, however, he sometimes suggested that "the miseries of subsisting by authorship" were probably worse, "To Bernard Barton [January 9, 1823]," in *Works*, 7:594.

21. Rajan, *Under Western Eyes*, 50.

22. Gregory Dart, *Metropolitan Art and Literature, 1810–1840: Cockney Adventures* (Cambridge: Cambridge University Press, 2012).

23. Lamb, "To Thomas Manning [December 25, 1815]," in *Works*, 6:481; while Lamb's late Elia essay on the "Barrenness of the Imaginative Faculty in the Productions of Modern Art" (1833) is critical of Martin's overwrought depiction of the demise of Belshazzar, it still relates the biblical story alongside an anecdote about the late monarch (as Prince Regent) and the Royal Pavilion at Brighton, Dart, *Metropolitan Art and Literature*, 173–174.

24. Bowen, *Business of Empire*, 149–150.

25. Catherine Hall and Sonya O. Rose, eds., *At Home with the Empire: Metropolitan Culture and the Imperial World* (Cambridge: Cambridge University Press, 2006), 3.

26. Charles Lamb, "The South Sea House," in *Elia and The Last Essays of Elia*, ed. Jonathan Bate (Oxford: Oxford University Press, 1987), 1 and 2. Further page references will be given in parentheses.

27. My emphasis. Elia's later reference to "soundings of the Bay of Panama!" (*Elia*, 2) alludes to another speculative bubble in the form of the Darien scheme of the late 1690s, the ill-fated Scottish project to colonize Panama.

28. Higgins, *Romantic Englishness*, 134; on proposals to make London more magnificent, see Holger Hoock, *Empires of the Imagination: Politics, War, and the Arts in the British World* (London: Profile, 2010), 361–372.

29. "Yes, thou must droop; thy Midas dream is o'er; / The golden tide of Commerce leaves thy shore," Barbauld's narrator states, "Eighteen Hundred and Eleven, a Poem," in Anna Letitia Barbauld, *Selected Poetry and Prose*, ed. William McCarthy and Elizabeth Kraft (Peterborough, ON: Broadview, 2002), 163.

30. Higgins, *Romantic Englishness*, 134; Gerald Monsman, *Confessions of a Prosaic Dreamer: Charles Lamb's Art of Autobiography* (Durham, NC: Duke University Press, 1984), 37. See also Simon P. Hull, *Charles Lamb, Elia and the London Magazine* (London: Pickering and Chatto, 2010), 62.

31. Monsman, *Confessions*, 38.

32. Lamb had a colleague at India House named William Evans; Samuel McKechnie, "Charles Lamb of the India House," *Notes and Queries* 191, no. 10 (November 16, 1946): 204–205.

33. Monsman, *Confessions*, 51; Simon P. Hull describes the clerks of the South Sea House as "eccentric, almost imbecilic," *Charles Lamb*, 62.

34. William Hazlitt, "On Londoners and Country People," in *Metropolitan Writings*, ed. Gregory Dart (Manchester: Carcanet, 2005), 85.

35. Dart, *Metropolitan Art and Literature*, 250–251; Dart notes that the action of volume 2 of *Great Expectations* is set in 1823.

36. Monsman, *Confessions*, 51; on newsmonger satire, see for example Brian Cowan, *The Social Life of Coffee: The Emergence of the British Coffeehouse* (New Haven, CT: Yale University Press, 2005), 235–243.

37. Karen Fang, "Empire, Coleridge, and Charles Lamb's Consumer Imagination," *Studies in English Literature, 1500–1900* 43 (2003), 834.

38. On contemporary attempts to negotiate this trauma, see for example Daniel O'Quinn, *Entertaining Crisis in the Atlantic Imperium, 1770–1790* (Baltimore, MD: Johns Hopkins University Press, 2011).

39. Dirks argues that "the greatest irony of this spectacular scandal was not that the trial fizzled out after nine long years, but that it led to the regeneration of the imperial idea"; Nicholas B. Dirks, *The Scandal of Empire: India and the Creation of Imperial Britain* (Cambridge, MA: Belknap Press of Harvard University Press, 2006), 85.

40. Lamb's friend Leigh Hunt wrote in 1818 that "the people in England seldom even think of India . . . India only presents itself occasionally to their minds, as a great distant place with strange beasts and trees in it, where Brahmins meditate and Musselmen keep seraglios,—where white people in regimentals are always fighting for some cause or other with the dark natives in vests and turbans,—and from which sallow elderly gentlemen are every now and then coming away to enjoy the large fortunes which they have acquired,—which they cannot do for the bile"; Leigh Hunt, "India," *The Examiner* 560 (September 20, 1818), 594.

41. It is difficult to be more specific than this, but Lamb's role in what H. V. Bowen terms "the continuing management of an expanding trade" (*Business of Empire*, 149–150) would probably have involved him recording imports of Chinese tea that had been exchanged for Indian-grown opium.

42. Higgins, *Romantic Englishness*, 134.

43. H. V. Bowen notes that "an ever-widening circle of people in Britain had been brought into economic contact with the Company since 1756," *Business of Empire*, 260–261.

44. "Prospectus," *The London Magazine* 1 (January 1820), iv. The phrase "mighty heart" is drawn from Wordsworth's sonnet "Composed upon Westminster Bridge (1802)"; it is worth noting here that the idea of "pulsations" emanating from the imperial metropolis and reverberating outward into the world sets up a unidirectional "center-periphery" model of the workings of empire.

45. Lamb, "To William Wordsworth [September 19, 1814]," in *Works*, 6:444. Contrast Lamb's writings in his "Londoner" persona, Dart, *Metropolitan Art and Literature*, 143–146.

46. Barbauld, "Epistle to Wilberforce, Esq. on the Rejection of the Bill for Abolishing the Slave Trade," in Barbauld, *Selected Poetry and Prose*, 126.

47. Joseph Addison, *Spectator* 69 (May 19, 1711), in Joseph Addison and Richard Steele, *The Spectator*, ed. Donald F. Bond (Oxford: Clarendon, 1965), 1:293.

48. Lamb, "To Thomas Manning [January 2, 1810]," in *Works*, 6:409.

49. Hazlitt, "The Indian Jugglers," in *Metropolitan Writings*, 3.

50. Daniel E. White, *From Little London to Little Bengal: Religion, Print and Modernity in Early British India, 1793–1835* (Baltimore, MD: Johns Hopkins University Press, 2013), 5.

51. Jon Klancher, "Discriminations, or Romantic Cosmopolitanisms in London," in *Romantic Metropolis: The Urban Scene of British Culture, 1780–1840*, ed. James Chandler and Kevin Gilmartin (Cambridge: Cambridge University Press, 2005), 67.

52. Monsman, *Confessions*, 42. Monsman reads Lamb's creation of his Elia persona in relation to the "moment of primal horror" (13) when his sister Mary stabbed their mother to death and seriously wounded their father.

53. Charles Dickens, *Great Expectations*, ed. David Trotter (Harmondsworth: Penguin, 1996), 264 and 480.

7

Commerce and Cosmology on Lord George Macartney's Embassy to China, 1792–1794

GREG CLINGHAM

1

The British embassy led by Lord George Macartney to the court of the Qianlong emperor in 1792–1794 is a pivotal event in the history of Sino-British relations. Historians have long seen the embassy as a diplomatic failure, for, as we know, Macartney was unable to negotiate the international trade agreement that both the East India Company and the British government sought. Nor was he able to persuade the Chinese to permit the British to establish a permanent diplomatic mission in Peking. Coming after Macartney's diplomatic successes in Russia and India, the failure of the China mission was a personal setback. Many historians and critics have seen Macartney's decision not to kowtow to the emperor in September 1793—in contravention of Qing protocol—as an imaginative as well as diplomatic failure. The accusations of failure as related to the kowtow began as early as the 1794–1795 Dutch embassy to China and have been repeated as recently as 2018 by the eminent historian Stephen R. Platt.[1] However, the traditional nature of Qing court culture under Qianlong, structured according to the cosmology of the Middle Kingdom, in which the godlike emperor stands at the center of the universe, in addition to its traditions of vassalage, gave Macartney's embassy—as well as Lord Amherst's in 1816—little chance of success in realizing their goals.[2] We know this not least because Qianlong's notorious letter to George III, putting the vassal nation in its place and declining the invitation to closer diplomatic and commercial ties, was written even before Macartney arrived in China.[3] Qianlong wrote: "We have never valued ingenious articles, nor do we have the slightest need of your country's manufactures. . . . As to what you have requested in your message, O King, namely to be allowed to send one of your subjects to reside in the Celestial Empire to look after your country's trade, this does not conform to the Celestial Empire's ceremonial system, and definitely cannot be done."[4]

It is not a neocolonial indulgence to propose that Macartney's behavior did not occasion the Chinese repulse, nor necessarily limit his effectiveness in dealing with the Chinese. Macartney's assumed heavy-handedness automatically forecloses close attention to his style of writing (and thinking). Yet it is precisely in his *writing*—in its nuances, ironies, and historical subtleties—where we glimpse Macartney's understanding of the cultural distance between the two worlds, and where we see his grasp of the cultural and commercial ties that *were already* shaping the world. In a concluding essay to his narrative of the embassy, Macartney acknowledges the limits of his—and any Westerner's—knowledge of China:

> The intercourse of the Chinese with foreigners is . . . so regulated and restrained, and the difficulty of obtaining correct information so great that the foregoing papers must not be received without reserve nor regarded otherwise than as merely the result of my own researches and reflections, for I am sensible that, besides being defective in many points, they will be found to differ a good deal from the accounts of former travelers.
>
> (MJ, 278)

His interest in bridging that cultural divide, however, expressed itself, among other things, in a cosmopolitan vision rooted in the economic principles of such Enlightenment economists and social thinkers as David Hume and Adam Smith, stressing the mutual advantages of trade.[5] Samuel Johnson articulates this cosmopolitanism in his preface to Richard Rolt's *New Dictionary of Trade and Commerce* (1756):

> The knowledge of trade is of so much importance to a maritime nation, that no labour can be thought great by which information may be obtained . . . there never was from the earliest ages a time in which trade so much engaged the attention of mankind, or commercial gain was sought with general emulation. . . . The merchant is now invited to every port, manufactures are established in all cities, and princes who just can view the sea from some single corner of their dominions, are enlarging harbours, erecting mercantile companies, and preparing to traffick in the remotest countries.[6]

In the case of Britain's trade with China, however, the benefits Johnson describes had not materialized: China had confined international trade to a single port, Canton, limited the trading season from October to March and only within licensed hongs, and they restricted the flow of bullion to Peking.[7] By the 1750s the British rage for chinoiserie, and the growing importation of Chinese goods—especially tea—made for a lopsided trade between the two countries. In the 1750s Hume noted, "China is represented as one of the most flourishing empires in the world, though it has very little commerce beyond its own territories."[8] Macartney verified this view when he visited Peking. He found a grand, populous city, "justly to be admired for its walls and gates, the distribution of its quarters, the width and allineation of its streets, the grandeur of its triumphal arches and the number and magnificence of its palaces." But trade was confined to two streets "that are chiefly inhabited by merchants and traders, whose shops and warehouses are most profusely decorated with every ornament that colours, gilding and varnish can bestow" (MJ, 156–158). Clearly, for Macartney the tinsel suggests the superficial, unsophisticated nature of Chinese society, in contrast to the rich textures and social complexities that characterized cities like London, that were benefiting from global commercial networks.

In an essay on Chinese "Trade and Commerce" supplemental to the narrative of the embassy, Macartney notes the ratio of Chinese trade with the East India Company as being 3:1 (MJ, 256–263),[9] with an even larger unofficial ratio in favor of China. No sinologist, Macartney nonetheless understands Chinese insularity. In an essay on Chinese "Arts and Sciences," he astutely notes the linkage between Chinese technology and population, and their attitude to trade: "Most of the things which the Chinese know they seem to have invented themselves, to have applied them solely to the purpose wanted, and to have never thought of improving or extending them further" (MJ, 267). While the implications of this situation do not bode well for British entrepreneurship, Macartney understands its historical logic:

it is true wisdom in the government to discountenance the use of the mechanic powers, or at least never to employ such artificial aid but where it is absolutely unavoidable, because the existence of so many millions of their people depends chiefly upon their manual labour.

(MJ, 267)

However, Hume and Smith equated concentration on local markets with cultural stagnation, and their idea of economics as the engine of social progress saw fiscal constraint not only as socially retrograde, but also as structurally flawed. For Smith, "the modern Chinese, it is known, hold [foreign commerce] in the utmost contempt, and scarce deign to afford it the decent protection of the laws."[10] Anticipating twentieth-century global economic circumstances, the Macartney embassy goes to China on the understanding that "foreign trade [is] every way confined within a much narrower circle than that to which it would naturally extend itself, if more freedom was allowed to it" (Smith, *Wealth of Nations*, 2:680). Macartney's attempt to broker a trade deal with the Chinese seeks to tap that vast economic reservoir.

His interest in Chinese culture, therefore, was highly purposeful, and being a sophisticated writer and experienced diplomat, he deployed his rhetorical skills strategically to further the commercial agenda of the East India Company, while also representing the achievements of his nation. In this essay, I consider how Macartney handles the gifts taken by the British to China as a kind of rhetoric, and I argue that his use of cosmological tropes is designed to achieve his commercial and diplomatic ends. At the same time, he was contributing to the commercial and cultural networks that were shaping Britain's geopolitical, economic, and cultural presence in the 1790s.[11]

2

Unpublished manuscripts in the Charles W. Wason Collection at Cornell University and the Toyo Bunko Oriental Library in Tokyo are central to my account.[12] These archives include letters and official memorandums passing between Macartney and the directors of the East India Company (at Cornell), and between Macartney and the ministers of the Qianlong emperor (in Tokyo). Among other things, these manuscripts reveal how much time and care the British committed to thinking about the gifts and how attuned they were to their symbolic importance at the Qing court.

The gift system of both commercial (Britain) and precommercial (China) societies clearly lent itself to symbolic manipulation in organizing legal and economic relations. Linda Zionkowski and Cynthia Klekar note that gift relations in eighteenth-century England manifested themselves most obviously in patronage. Through the logic of gift exchange, they remark, assertions of power in the form of patronage by the ruling orders were acts of service for which deference was the necessary return.[13] This gift economy in contrast to the market is, in Pierre Bourdieu's words, "organized with a view to the accumulation of *symbolic* capital," and it ensures the stability of the social hierarchy. The manuscripts at Cornell and the Toyo Bunko show that Macartney understood very well this imbrication of commerce with gifting. He saw, in Bourdieu's words, that the "rhetoric of gift giving was employed to advance the objectives of this society, *including* the circulation of commodities (both inanimate and human) necessary for the growth and reproduction of an international commercial culture."[14] A difficulty arose for Macartney, however,[15] because gifting was part of Chinese cosmology, which identified the relations of other nations to China as implying subservience, rather than difference or equality. Eighteenth-century European aristocratic culture would perhaps have seen any subservience implied by the transaction of giving and receiving, as attached to the *recipient*, rather than to the giver, but not the Qing. Macartney was less knowledgeable about Chinese manners and protocols than the sinologists on the later Amherst embassy—Sir George Thomas Staunton and Rev. Robert Morrison—and he was certainly unable to disentangle the discursive subtleties of the Qing in order to bridge the cultural divide between the two cultures.[16]

As early as his Russia posting (1764–1768), Macartney saw the world as a series of potential relationships rather than as opposition. He pursued diplomacy-as-economic-cooperation rather than diplomacy-as-balance-of-power, in the tradition of the French theorist Jacques Turgot and the American pragmatist John Adam.[17] However, it is as a *writer*, in his handling of language and narrative representation that Macartney truly demonstrates his diplomatic skills. Certainly, he knew that the Chinese saw themselves as superior to other nations—having even made notes to this effect when conversing in St. Petersburg in 1764 with a Russian by the name of Bratishchev who had been an official in Irkutsk.[18] Yet in China, Qing officials appropriated the purpose of the British embassy, treating it as a tribute to the Qianlong emperor on his birthday, for whom there were extensive festivities, attracting representatives from many vassal nations.

While Macartney was willing to participate in the emperor's birthday celebrations as a *tributary guest*, he was *unwilling* to subordinate himself—and the nation he represented—in his diplomatic negotiations, where he expected to be an equal. As already mentioned, most accounts of this embassy treat Macartney's expectation of diplomatic equality as a flaw, and his refusal to kowtow to the Qianlong emperor as arising from ignorance or ethnocentricity. Alain Peyrefitte's detailed retelling of the story of the embassy is based on the proposition that "an apparently trivial incident"—the refusal to kowtow—"symbolized Macartney's failure."[19] Peter Kitson, more recently, notes, "Macartney shows no serious understanding of the semiotics of Qing guest ritual [and] presents his refusal to kowtow as a success and an instance of the benefits of British firmness and rectitude."[20] Lydia Liu, taking up a clear postcolonial stance, correctly argues that "the issue of the *koutou* (kowtow) and other forms of humiliation during Lord Macartney's mission to the Qianlong emperor's court obsessed the British imagination for nearly two centuries," which for her explains the British humiliation of China throughout the nineteenth century.[21] There may have been, as Liu says, a "startling level of neurotic fixation that had driven the British ruling class to engage in one military campaign after another in China" (217), but this was not Macartney's fixation, nor his humiliation, nor his military campaigns. To read Liu's historiographical interpretations of nineteenth- and twentieth-century events *back* into 1793 is deliberately ahistorical and deprives Macartney of the agency that is obviously present in his narrative, and, to some extent, in his engagements with the Chinese.

A letter in the Wason Collection, dated February 20, 1792, from John Ewart, a civil servant, to Henry Dundas, then Minister for the Colonies, accompanying a feasibility study for the upcoming embassy ("Facts & Considerations relative to the proposed Embassy to China") (III: docs. 85–86), entertains the probability that Macartney *would be* expected to kowtow to the emperor. These are fascinating, informed documents, worth studying for what they reveal of British political culture, as are Macartney's equally detailed, informed, and shrewd responses. Ewart writes as follows:

It has been usual for the Nations desirous of forming an alliance with the Chinese who are themselves so much devoted to Parade, to fit out their Embassies with singular magnificence and splendor, in order to impress that supercilious Nation with ideas of grandeur equal to their own. No Embassies are omitted to the Emperor, which do not bring valuable or curious presents; & the Emperor makes a suitable return to the Sovereign who sends them. The Emperor of China requires likewise marks of submission, not conceded by the Embassadors of a Sovereign to any other Prince, such as prostrating before Him and beating the forehead 9 times on the floor; and without these Condescensions no Embassador was ever received. They are nevertheless considered more as a form than a submission, for the Emperor's Prime Minister returns the same marks of reverence to the credential letter presented by the Embassador to the Sovereign. (Wason Collection, III: doc. 85, fols. 45–46)

On March 17, 1792, Macartney responded at length to Ewart's document, including the following passage on the kowtow:

As to the genuflexions, prostrations and other idle oriental ceremonies mentioned in this Section, such matters are always to be arranged by discussions, explanations and mezzo [blank] according to the discretion of the Embassador and Characters of those whom he has to treat with.
                    (Wason Collection, III: doc. 95, fols.7–8)

Simultaneously, Macartney sent Dundas a copy of Johan Nieuhof's *An Embassy from the East-India Company of the United Provinces, to the Grand Tartar Cham, Emperour of China* (1669).[22] This account of an embassy is notable for its empirical attempt to demystify China, and for the fact that the Dutch unproblematically accepted the need to kowtow to the Shunzhi emperor, the third Qing emperor, helping to secure extensive trading privileges.

The expectation of the kowtow thus did not come as a surprise to Macartney, nor should his decision to attempt to "arrange" the matter by discussion and explanation. For the kowtow for Macartney was neither a personal nor an ideological issue. He had no philosophical objection to the three Chinese rites that Voltaire highlights in his account of the reign of Louis XIV (1751)—the name for the deity, ancestor worship, and the cult of Confucius. What was at issue for Macartney was neither dominance nor "imperial ambition," as Liu would have it (58), nor "masculine" values, as in Kitson's account (a term used by Staunton and Barrow, but not by Macartney). *He* was interested in reciprocity and recognition. If read with attention, his text—both published and unpublished—is quite clear that he is engaged in a *symbolic* transaction that seeks to accord Qianlong and George III the same treatment and value.

> They asked me what was the ceremony of presentation to the King of England. I told them it was performed by kneeling upon one knee and kissing His Majesty's hand. "Why then," cried they, "can't you do so to the Emperor?" "Most readily," said I; "the same ceremony I perform to my own King I am willing to go through for your Emperor, and I think it is a greater compliment than any other I can pay him."
>
> (MJ, 119)

This is part of Macartney's diplomatic effort to see political relations as a cultural network. His negotiation over the kowtow is an attempt to *trade* with the Chinese, for he is asking the Chinese to give up something of value in exchange for the acquisition of another of value. For Macartney to kneel to the emperor, while one of the Chinese ministers kneeled to a portrait of King George III (they took with them Reynolds's 1779 portraits of the king and of Queen Charlotte),[23] was an act of *symbolic* thinking, whereby Qianlong and George III are equated with each other, and both are incorporated into the larger network. That this network presupposed Britain's imperial and commercial ambitions is unsurprising. Writing in 1782 from India to Charles Jenkinson, Secretary of War, Macartney declared, "I really feel so interested in this immense stake of the Empire, that I would make any sacrifice to preserve it for I know its Importance."[24] However, that was India not China, where the remit was wholly different. In China, Macartney's writing and thinking repeatedly avoid the ideological and aim instead for the practical and the historical.

Still, when the Chinese officials accept Macartney's suggested mode of salutation, they insisted on omitting the kissing of the hand, and they *cut* the reciprocity altogether by which the Chinese minister kneels to the portrait of the king. The British and the Chinese were of course dependent on translators for their communication, and it is possible, perhaps likely, that nuanced meanings and subtle intentions were lost in translation.[25] Ho-shen, when writing for the emperor to other ministers, was uncompromising in his disapproval:

> we are extremely displeased at his unwarranted haughtiness, and
> have given instructions to cut down their supplies . . . if, when
> the barbarians come for an audience with the Emperor, they are
> sincere and reverential, we always grant them our favour, so as
> to display our "cherishing kindness." If they tend to be in the
> least haughty then they are not destined to receive our favour.[26]

Macartney, however, was baffled: to omit half the ceremony on which they had agreed was to undermine its symbolic coherence. He reflects: "thus ended this curious negotiation, which has given me a tolerable insight into the character of this Court, and that political address upon which they so much value themselves" (MJ, 119).[27]

3

The manuscript correspondence at Cornell between Macartney and Henry Dundas and between Macartney and the king reveals how highly they all thought of the Chinese. As Macartney writes to Dundas on January 4, 1792, China is "the most civilized as well as most ancient and populous Nation on the Globe," it has "celebrated institutions," and it can bestow "benefits which must result from an unreserved and friendly intercourse" with other nations (Wason Collection, II: doc. 28, fol. 1r). British aims were carefully framed in both cultural and commercial terms: "such a Philosophical Spirit might easily ally Itself to and greatly assist the progress of mercantile arrangements, of which the object to supply the mutual wants and increase the happiness of mankind is itself included among the Principles of soundest Philosophy" (fol. 2v).

This optimistic vision of the benefits of trade, extending across international boundaries and oceans—reminiscent of Hume and Smith—puts a philosophical gloss on the ambitions of the East India Company, which were quite complacent in contrast to Macartney's measured realism. For example, Francis Baring and John Smith Burges (chairman and deputy chairman of the company, respectively) wrote at one point during the preparations: "It is . . . evident that the Chinese are disposed to facilitate our views as much as possible, by promoting the favorite object of the Company" (Wason Collection, II: doc. 27, fol. 4). In private Macartney was not past indulging in a similar chauvinism: his unpublished China commonplace book contains comments such as the following (Figure 7.1):

> The Chinese (Tartars included) are certainly impresst with a high idea of the English from this Embassy, & not a little mortify'd, I believe, from the vast superiority they perceive in us. They were surprized, I understand, to find us rich, sociable, good humor'd, & not presuming, tho so much enlightened & so capable & ready to teach them, many things that they feel they want of & wish to know, tho impeded by their laws & usages, which, regulate their tastes, manners & behavior—[28]

Notwithstanding such private indulgences, Macartney knew that in dealing with the Chinese he would have to resort to what he called "that dexterity and intrigue, which it may be sometimes thought useful to have recourse to at other Courts" (Wason

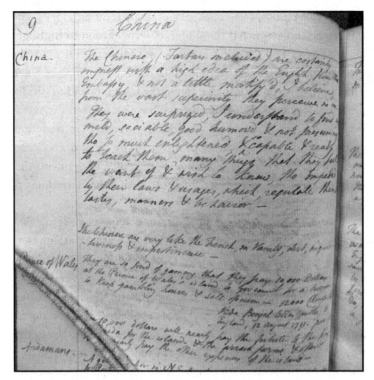

FIGURE 7.1. Entry in Sir George Macartney's Commonplace Book kept during his Embassy in China, 1792. Cornell University Library, MSS DS M118, fol. 9. Photograph by the author.

Collection, II, doc. 27, fol. 3v). The gifts taken by the embassy for the emperor and his officials were, in a sense, instruments of dexterity and intrigue.

The embassy carried with them samples from many aspects of British art, science, technology, manufacturing, and war. The Wason Collection includes forty-one manuscript folios enumerating the "gifts taken to China, with explanations for their assemblage, & costs" from the East India Company (V: fol. 225); and among other documents, it also includes a catalog of presents sent by the king with explanations in English and Latin running to twelve folios (VIII: fol. 350). The British filled one entire ship with paintings, porcelain, vases, tapestry, carpets, manufactures, furniture, weapons, telescopes, globes, planetariums, clocks, and a Herschel reflector. All these objects were chosen to highlight British artistic and scientific invention and cultural greatness—the "vast superiority" Macartney mentions in his commonplace book. The purpose was not only to impress the Chinese, and to display philosophical and scientific principles, but also, apparently, to *persuade* the Qing court to adopt British proposals for trade and communication. In this array of merchandise most of the items highlighted by Macartney pertain to temporality, instruments that measure time, or space-time, or timing, and the harmonies of astronomical understanding.

This choice was strategic, because in placing clocks, orreries, and globes at the center of his narrative Macartney attempts to create a scene, an ethos, in which the two nations can engage in sociable dialogue while locating themselves in universal time. While Qing cosmology placed the emperor and the Middle Kingdom at the center of an eternal, static universe, Enlightenment astronomy was developing theories of flux and change. For example, Sir William Herschel (1738–1822), musician, composer, conductor, and astronomer royal (1782), used telescopes to expand the knowledge of the universe and the galactic system or nebulae as a vast network of stars.[29] Herschel's paper, "On the Construction of the Heavens" (1785), in the *Philosophical Transactions*, described an important astronomical discovery that would indirectly influence Macartney, for, as Michael Hoskin notes, "it privileges the role of gravity in the evolution of the universe of stars."[30] In drawing on Enlightenment cosmology to influence Qing social and psychological experience, Macartney took with him not only ideas, but also a Herschel reflector used in the "discovery" of Uranus (1781), its two satellites (1787), and Saturn's two moons (1789).

4

Matteo Ricci (1552–1610) and his teacher Michele Ruggieri (1543–1607) were responsible for introducing Western clocks into China as a way of using science to promote Christianity.[31] Catherine Pagani recounts how, in his attention to clocks, Macartney was attempting to tap into Qianlong's admiration for that particular Western technology.[32] Qianlong's magnificent collection of European timepieces and astronomical devices, largely inherited from his grandfather, the Kangxi emperor, were considered as an expression of the status and power of the Qing, even while Qianlong identified them as Western.[33] During Qianlong's reign, clocks were for the first time included in court documents, and they featured in his own writings and in his son's, the Jiaqing emperor.[34] Macartney, consequently, is very attentive to clocks and astronomical devices in his published narrative—their identities, their transportation to Peking, their assemblage, their placement, their presentation to the emperor, and who would eventually see and learn from them. In an extraordinary fantasy of cultural appropriation, if not imperial aggression, Macartney imagines the gifts as occupying a position at the center of the Yuen-min-Yuen, the Emperor's palace (Figure 7.2).

FIGURE 7.2. "Hall of Audiences at Yuen-min-Yuen." From Sir George Leonard Staunton, *An Authentic Account of an Embassy from the King of Great Britain to the Emperor of China* (London, 1796–1797), volume 3, plate 22, no pagination. Photograph by the author.

On one side of the throne was to be placed the terrestrial globe, on the other the celestial; the lusters were to be hung from the ceiling, equal distances from the middle of the room; at the north end the planetarium was to stand; at the south end Vulliamy's clocks, with the barometer and Derbyshire porcelain vases and figures, and Fraser's Orrery; an assemblage of such ingenuity, utility and beauty as is not to be seen collected together in any other apartment, I believe, of the whole world besides.

(MJ, 96)

Desirous to highlight their gifts in the most advantageous way, among the thousands of other foreign tribute items, Sir George Staunton, in *his* account of the embassy, notes:

It was necessary to make out, somewhat in the Oriental style, such a general description of the nature of the articles . . . as appeared likely to render them acceptable; measuring their merit by their utility, and endeavouring even to derive some credit from the omission of splendid trifles.[35]

Alain Peyrefitte thinks this a "bombastic" idea that overlooked the first principle of Chinese courtesy: the donor should understate the value of his gifts, in order not to humiliate the recipient. Since Macartney did not characterize his presents as "little nothings," mere "trinkets from our poor country," he wound up sounding arrogant.[36] However, Staunton's reportage may convey overconscientiousness, but there is no arrogance, and neither is there any prima facie arrogance in Macartney's detailed account of the gifts either in the published narrative or in the Toyo Bunko manuscripts. Something essential may again have been lost in (cultural) translation. In the preface (fol. 63) to the annotated Catalogue of Presents in the Toyo Bunko, Macartney includes painstaking directions and explanations. These manuscripts show us Macartney articulating a *cosmological* framework within which he wished the Chinese to *think about* the clocks and the astronomical instruments. His narrative seeks to reconcile several demands: to appeal to the emperor's eminence, to exemplify the ingenuity of British technology, to present it as a manifestation of the British monarch—equal to but different from Qianlong—and to persuade the emperor to join them in a commercial transaction in the name of a philosophical enterprise.

These objects, Macartney writes, "may denote the Progress of the Sciences and of the arts in Europe—and which *may convey some kind of information* to the exalted mind of his Imperial Majesty, or such others as may be practically useful. The *intent not the presents themselves,* is of value between Sovereigns" (my emphasis). Some form of symbolic reading would thus be required to appreciate and to respond to Macartney's account, as he seeks to link cosmology, science, culture, and commerce. He starts with the planetarium built by William Fraser, Royal optical and mathematical instrument maker.[37] This planetarium consists:

> Of many parts, which may be distinctly used or connected together, [it] includes the whole universe of which the Earth is but a small proportionate part. This vast machine is the utmost effort of Astronomical science and mechanic art, combined together that was ever made in Europe. It shows and imitates with great clearness and mathematical exactness the several motions of the Earth according to the System of European Astronomers, likewise the eccentric or irregular motions of the moon around it, and also the stations, and movements of the Stars; and of the Sun and Planets which surround it.

We note here that Macartney's unfolding cosmology not only sees the earth as a small and insignificant part of the universe, but also identifies this *idea* as being the work of "European Astronomers." This would seem to privilege European science at the expense of Chinese lore. However, as he continues Macartney seeks to reposition Qianlong at the center of historical and social attention, if not of the universe.

Another part indicates the month, the week, the day, the hour and minute at the time it is looked at. It is calculated for more than a thousand years; and it is as simple in its construction as it is complicated and wonderful in its effects. No such machine remains behind in Europe and for above a thousand years it will be a monument of the respect in which the virtues of His Imperial Majesty are held in the remotest parts of the world.

Images of Fraser's planetarium are not readily available, but other contemporary planetariums and orreries, made by Benjamin Martin (1704–1782),[38] lexicographer and scientific instrument maker, with the prodigious John Senex (bap. 1678, d. 1740),[39] cartographer and engraver, now at Harvard University, give a good sense of what Macartney presented to Qianlong (Figures 7.3, 7.4).

Martin's philosophical idea of the Orrery stands behind Macartney's enthusiastic account of Fraser's planetarium. The orrery is a "bare Representation of the Solar System in its native

FIGURE 7.3. Large orrery made by Benjamin Martin in London, 1766, used by John Winthrop to teach astronomy at Harvard, on display at the Putnam Gallery, Harvard University. Photograph by Sage Ross (November 24, 2009), from Wikimedia Commons.

FIGURE 7.4. Orrery made by Benjamin Martin in London, 1767, used by John Winthrop to teach astronomy at Harvard, on display at the Putnam Gallery, Harvard University. Photograph by Sage Ross (November 24, 2009), from Wikimedia Commons.

Simplicity" and it is "no less than a universal Representation of the Mundane System, and a Explication thereby of all its numerous Phenomena."[40] A painting such as Joseph Wright of Derby's *A Philosopher Giving That Lecture on the Orrery . . .* (1766) (Figure 7.5) captures the symbolic resonance of the orrery in eighteenth-century culture, symbolizing, as Jan Golinski says of astronomy in general, "the triumphs of the scientific method," having "both observational and theoretical dimensions."[41]

Wright's famous oil painting depicts a public lecture about a model solar system, with a lamp—as the sun—illuminating the faces of the audience. Wright captures the spirit of Enlightenment scientific inquiry as a moral and sociable force for the audience of ordinary folk and learned men under the instruction of a natural philosopher. For Benedict Nicolson, this is a moment when "the mystery of science was beginning to weave its thread into everyday life and play its part there as common sense, while still

FIGURE 7.5. Joseph Wright of Derby, *A Philosopher Giving That Lecture on the Orrery, in Which a Lamp Is Put in Place of the Sun*, circa 1766, oil on canvas, 147.2 × 203.2 cm, Derby Museum and Art Gallery. Photograph by europeana.eu from Wikimedia Commons.

retaining, on its way to sense, the magic of the unknown."[42] David Solkin, in turn, argues that Wright uses a scientific experiment not only to celebrate astronomy, but also to "instruct his audience about the character of their own world," thus embracing a philosophical agenda for the visual arts in the vein of the Earl of Shaftesbury and George Turnbull.[43] As we share the individuals' wonder and interest, we respond to Wright's narrative of cultural progression that, as does Macartney, links astronomy to sociability and humanity. There is no record of Macartney's knowledge of Wright, yet the narrative of his engagement with the Chinese attempts to translate the aesthetics *of* a Joseph Wright into rhetorical form. Whether or not Qianlong and his ministers were remotely inclined to read astronomical instruments in this spirit of hermeneutic inquiry, Macartney sought to persuade them to think symbolically about the orrery. For Macartney, as for Martin describing Thomas Wright's Great Orrery of 1730 (Figure 7.6), the orrery is an "adequate representation of a TRUE SOLAR

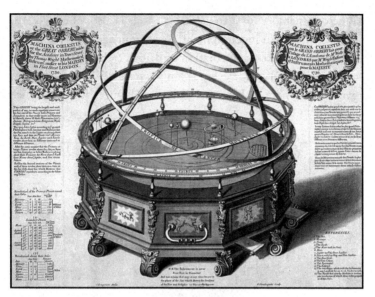

FIGURE 7.6. Thomas Wright, *Machina Coelestis, or the Great Orrery*, 1730. Engraved by G. Vandergucht after C. Lemprière. 20 × 28 inches. Science Museum Group Collection, London.

SYSTEM, [which] gives a just Idea of the *Number, Motions, Order*, and *Positions* of the heavenly Bodies."[44]

As Jesse Molesworth argues, the orrery as "adequate representation" becomes a "just idea" by shifting the focus from magnitude to movement, from distance to motion, which, crucially for Macartney, enabled a representation of time. This was not clock time, but cosmological time, the idea that celestial objects orbited the Sun within predictable timeframes, thus enabling the observer to place particular moments of chronological time within the continuum of cosmic time.[45]

Within such a context of cosmological associations, Macartney recommends Fraser's planetarium to Qianlong as a machine that represents the emperor's universal, quasi-divine status at the center of the Middle Kingdom. To enhance this notion, Macartney points out that Fraser's orrery is unique—he says, "No such machine remains behind in Europe"—and that it shall preserve the emperor's memory for a thousand years. It is also, of course, the product of British ingenuity, and its assemblage, operation, and explication require the mediation of British technicians.

Fraser's orrery, however, is only *supplementary* to another invention of note, a Herschel reflector (Figures 7.7, 7.8), which

FIGURE 7.7. Friedrich Wilhelm Herschel, model of telescope used in discovery of Uranus, 1781. 7 feet long, 6 inches diameter, *f*14 reflector. Herschel Museum of Astronomy, Bath, United Kingdom. Photograph by Mike Young via Wikimedia Commons.

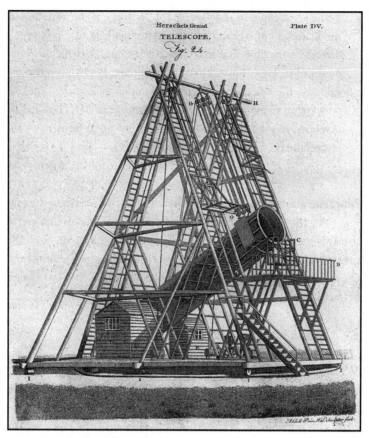

FIGURE 7.8. Friedrich Wilhelm Herschel, reflector telescope, 1789. 40-foot focal length, 48-inch diameter primary mirror. Illustration in *Encyclopaedia Britannica*, 3rd ed. (Edinburgh: A. Bell and C. MacFarquhar, 1797), vol. 18. From Wikimedia Commons.

illuminates the phenomena modeled by the planetarium. While the planetarium is an instrument that *imagines and represents* the organization and operation of the solar system, the reflector provides direct access to the vastness of the universe itself. This, according to Macartney, is a more "curious and useful" object:

> observing farther and better than had hitherto been done the most distant and minute heavenly bodies, and they really move

in the great expanse and *proving* the accuracy of the imitation in the Planetarium. These observations are made not by looking directly at the object, as in common telescopes, in which the power of seeing is limited to certain degrees, but by perceiving sideways the reflection of such objects upon mirrors. . . . The power[s] of vision are extended by the means of Mr. Herschel's machine, beyond the hope, or calculations of all former philosophers.

(Toyo Bunko, fol. 63, article 1)

Macartney concludes the description of Herschel's reflector by saying: "As the Study of astronomy is not only essentially useful towards the perfection of Geography and navigation, but from the greatness of the objects to which it relates, it elevates the mind and thus is worthy of the contemplation of Sovereigns, and has always attracted the notice of His Imperial Majesty."[46]

5

We might draw the following conclusions from Macartney's orchestration and explication of the gifts to the Qianlong emperor. According to cosmographer Benjamin Martin, the orrery measures the relationship between human time and the cosmos. The cosmology inscribed in Macartney's account of Fraser's planetarium thus functions to place the king of Great Britain and the emperor of China harmoniously together in a kind of universal center. Macartney takes *literally* the ancient belief of the Mandate of Heaven (天命; pinyin: tiānmìng; literally: "heaven named") as applied to Qianlong; he then fictionalizes it for his own purpose. The power of this fiction, however, depends on the emperor accepting the terms in which it is offered; that is, to read both events and technology symbolically, as pertaining to both the real commercial world and to Qing myth. While the planetarium provides a fictional model of the solar system, the Herschel reflectors allow us to see the solar system in operation. As Macartney points out to the emperor, the study of astronomy elevates the mind and is "worthy of the contemplation of Sovereigns." To enter the fictional world created by Macartney's rhetoric—which turns out also to be the real world as uncovered by British astronomy—*is* to participate in the highest purposes of sovereignty.

Macartney's diplomatic efforts failed; but to regard his handling of gift exchange as a failure because it did not achieve its goal of eliciting a trade agreement *is* itself a failure of historical imagination. Suggestive as is Macartney's published narrative with regard to his thinking about gifts, the manuscripts at Cornell and the Toyo Bunko lift the veil on Macartney's thought and reveal his subtle and sophisticated linking of cosmology with commerce in the British engagement with China in 1792–1794.

Notes

A short version of this essay was delivered as a lecture at a conference, "The Inexplicable and the Unfathomable: China and Britain, 1600–1900," at the Courtauld Institute of Art, London, November 2016. My thanks to Professor David Park of the Courtauld Institute for the invitation. My work on Macartney in China was, to a large extent, stimulated by the example of my Bucknell colleague, the distinguished scholar of Chinese, the late James R. Pusey. I am deeply grateful to Jamie for his encouragement and friendship.

1. See Stephen R. Platt, *Imperial Twilight: The Opium War and the End of China's Last Golden Age* (New York: Vintage, 2018), 35–36. A Dutch American merchant on a 1794–1795 Dutch embassy to China claimed to be reporting Chinese views on the matter of Macartney's refusal to kowtow; see Andreas Everardus van Braam Houckgeest, *Voyage de l'Ambassade de la Compagnie des Indes Orientales Hollandaises, vers l'Empereur de la Chine, dans les années 1794 et 1795, publié en français par M. L. E. Moreau de Saint-Méry. Philadelphie* (Philadelphia, 1797), 2:v–vii. My thanks to Professor Roderick Whitfield of the School of Oriental and African Studies, London, for the van Braam reference. An English version of van Braam's journal appeared in 1798: *An Authentic Account of the Embassy of the Dutch East-India Company, to the Court of the Emperor of China, in the years 1794 and 1795* (London:

R. Philips, 1798). The English version has completely different notes from the original French. Van Braam begins with A for "Ambassade," then goes on to C for "Cours de Justice." The English starts with B ("bamboo," "Barrow"), and has no "courts of justice" under C. Could the publisher have been censoring the criticism of the Macartney embassy in the year after Sir George Staunton's "official" account of the Macartney embassy appeared?

2. For the Amherst embassy, see Peter J. Kitson and Robert Markley, eds., *Writing China: Essays on the Amherst Embassy (1816) and Sino-British Cultural Relations* (Cambridge: D. S. Brewer, 2016), especially Kitson (56–82) and Markley (83–104).

3. See J. L. Cranmer-Byng, "Lord Macartney's Embassy to Peking in 1793, from Official Chinese Documents," *Journal of Oriental Studies* 4, no. 1–2 (1957–1958), 133. Cranmer-Byng demonstrates that the letter was drafted by Ho-shen, one of the emperor's ministers, on August 3, 1793, formally proclaimed on September 23, and handed to Macartney on October 3.

4. "An Edict from the Emperor Ch'ien-Lung to King George the Third of England," in Sir George Macartney, *An Embassy to China: Being the Journal Kept by Lord Macartney during His Embassy to the Emperor Ch'ien-lung, 1793–1794*, ed. with introduction and notes by J. L. Cranmer-Byng (London: Longmans, 1962), 340, 338. This edition of Macartney's embassy narrative and supplementary essays cited as MJ.

5. See Greg Clingham, "Cultural Difference in George Macartney's *An Embassy to China, 1792–94*," *Eighteenth-Century Life* 39, no. 1 (2015), 1–29.

6. Samuel Johnson, *The Yale Edition of the Works of Samuel Johnson*, vol. 20, ed. O M Brack, Jr., and Robert DeMaria, Jr., *Johnson on Demand: Reviews, Preface, and Ghost-Writings*, "Preface to Richard Rolt, A New Dictionary of Trade and Commerce (1756)" (New Haven, CT: Yale University Press, 2018), 245–252.

7. For Johnson's extensive cultural engagement with China, see Greg Clingham, "Johnson and China: Culture, Commerce, and the Dream of the Orient in Mid-Eighteenth-Century England," *1650–1850: Ideas, Aesthetics, and Inquiries in the Early Modern Era* 24 (2019), 178–242.

8. David Hume, *Essays: Moral, Political and Literary* (Oxford: Oxford University Press, 1963), 271.

9. Platt (*Imperial Twilight*, 47–48) asserts that Macartney's essays supplemental to the embassy narrative were written in anger to rationalize his failure, but the intellectual comprehensiveness, the very wide array of information, the reasonable tone, the historical and anthropological interest of these essays—which cover many topics, including manners and character, religion, government, law and justice, property, population, the economy, the military, trade and commerce, arts and sciences, technology and navigation, and the Chinese language—tell a completely different story. The essays are to be found between pages 221 and 278 of Cranmer-Byng's edition of Macartney's text used for this essay.

10. Adam Smith, *An Inquiry into the Nature and Causes of the Wealth of Nations*, ed. R. H. Campbell, A. S. Skinner, and W. B. Todd (Oxford: Clarendon, 1982), 1:495.

11. Recent helpful scholarship discussing the cultural and imaginative networks between Britain and China during the eighteenth- and early-nineteenth centuries include David Porter, *Ideographia: The Chinese Cypher in Early Modern Europe* (Stanford, CA: Stanford University Press, 2001), Chi-ming Yang, *Performing China: Virtue, Commerce, and Orientalism in Eighteenth-Century England* (Baltimore, MD: Johns Hopkins University Press, 2011), Eugenia Zuroski Jenkins, *A Taste for China: English Subjectivity and the Prehistory of Orientalism* (Oxford: Oxford University Press, 2013), and Peter J. Kitson, *Forging Romantic China: Sino-British Cultural Exchange, 1760–1840* (Cambridge: Cambridge University Press, 2013).

12. Original manuscripts, papers, and letters relating to the Macartney mission to China 1792–1794, MSS DS117; Charles W. Wason Asia Collections, Kroch Rare Book and Manuscript Library, Cornell University, Ithaca, NY. The Macartney papers in the Wason Collection consist of ten bound folio volumes, and I cite material from these manuscripts by volume number, document number, and folio. The Japanese archive is the Toyo Bunko Oriental Library, Tokyo, founded in 1924, incorporating the book and manuscript collection of the Australian sinologist and journalist G. E. Morrison (1862–1920). The library's holdings include several embassy-related printed materials and manuscripts, including one of two fair copies of Macartney's narrative,

the other being in the Beinecke Rare Book and Manuscript Library at Yale. The documents in the Toyo Bunko on which I draw in this essay are 166 loose papers in twelve envelopes entitled simply "Macartney Papers," call number 貴重書 MS-42.

13. See Linda Zionkowski and Cynthia Klekar, "Introduction" in *The Culture of the Gift in Eighteenth-Century England*, ed. Linda Zionkowski and Cynthia Klekar (New York: Palgrave, 2009), 1–12.

14. Bourdieu quoted by Zionkowski and Klekar, "Introduction," 4.

15. Not, as Klekar argues, because he failed to understand the relationship between gifting and commerce: see Cynthia Klekar, "'Prisoners in Silken Bonds': Obligation, Trade, and Diplomacy in English Voyages to Japan and China," *Journal for Early Modern Cultural Studies* 6, no. 1 (2006), 84–105.

16. See, for example, James L. Hevia, *Cherishing Men from Afar: Qing Guest Ritual and the Macartney Embassy of 1793* (Durham, NC: Duke University Press, 1999).

17. See Felix Gilbert, "The New Diplomacy of the Eighteenth Century," *World Politics* 4, no. 1 (October 1951), 1–38.

18. See Alain Peyrefitte, *The Immobile Empire* (New York: Vintage Books, 1992), 31–32.

19. Alain Peyrefitte, *The Collision of Two Civilisations: The British Expedition to China 1792–4*, trans. Jon Rothschild (London: Harvill, 1992), xviii.

20. Peter J. Kitson, "'Bid him bow down to that which is above him': The 'Kowtow Controversy' and the Representations of Asian Ceremonials in Romantic Literature," in *Coleridge, Romanticism and the Orient: Cultural Negotiations*, ed. David Vallins, Kaz Oishi, and Seamus Perry (London: Bloomsbury, 2013), 22, 21

21. Lydia H. Liu, *The Clash of Empires: The Invention of China in Modern World Making* (Cambridge, MA: Harvard University Press, 2004), 57.

22. Macartney writes to Dundas: "I beg leave to send you herewith for your inspection the Journal of the Dutch Embassadors in 1655, whose mission took place but a few years after the conquest of China by the Tartars in 1640, during a time of commotion and Jealousy, when the new Government had not as yet assumed a stable form or consistence, which circumstances may in a great measure account for the incomplete success of that Embassy. The Journal will however serve to shew a

variety of curious matters, which still subsist at this day, nearly as they did formerly" (Wason Collection, III: doc. 95, 10–11). The original Dutch work, *Het Gezantschap der Neêrlandtsche Oost-Indische Compagnie aan den Grooten Tartarischen Cham, den Tegenwoordigen Keizer van China* (Amsterdam, 1665), was unusual for illustrating its text with 149 engravings that supported its aim of representing real life in China.

23. See the Royal Academy website for these paintings: https://www.royal academy.org.uk/art-artists/work-of-art/portrait-of-king-george-iii.

24. Sir George Macartney, *The Private Correspondence of Lord Macartney Governor of Madras (1781–85)*, ed. C. Collin Davies (London: Royal Historical Society, 1950), 186.

25. In writing of the negotiations about the kowtow, J. L. Cranmer-Byng remarks: "It would be interesting to know exactly what Ho-shen said in Chinese and how Jacobus Li translated it into Italian. Macartney, perhaps, received too optimistic an idea of what Ho-shen actually intended"; in Peter Roebuck, ed., *Macartney of Lisanoure 1737–1806: Essays in Biography* (Belfast: Ulster Historical Foundation, 1983), 232.

26. Quoted by Cranmer-Byng from Chinese documents, see "Lord Macartney's Embassy," 158–159.

27. Platt cites the notebook of the twelve-year-old George Thomas Staunton (who accompanied his father, George Leonard Staunton, on the Macartney embassy) and the testimony of the Jiaqing emperor in 1816, during the Amherst embassy to China, to suggest that Macartney probably did kowtow to Qianlong when they met but that the fact was kept quiet (*Imperial Twilight*, 168–169). But there is nothing to even hint at this idea in Macartney's writings, published or unpublished, nor, I suggest, would he have had any difficulty in acknowledging a kowtow had it occurred, given the pragmatic cast of his diplomacy and the criticism he attracted for *not* doing so.

28. "Commonplace Book kept by Lord Macartney during his Embassy in China 1792," MSS. DS. M118, fol. 9, Charles W. Wason Asia Collections, Kroch Rare Book and Manuscript Library, Cornell University, Ithaca, NY.

29. See, for example, Michael Hoskin, *The Construction of the Heavens: William Herschel's Cosmology* (Cambridge: Cambridge University Press, 2012), chapters 4 and 5 (54–75). Hoskin includes Herschel's cosmological

papers from *Philosophical Transactions* published between 1783 and 1814, including "On the Construction of the Heavens" (1785) (113–135). See also Hoskin, *William and Caroline Herschel: Pioneers in Late 18th-Century Astronomy* (London: Springer, 2014). Herschel's sister, Caroline (1750–1848), collaborated in making these discoveries, also in writing up their research for publication, while pursuing her own independent astronomical work. See Marilyn B. Ogilvie, *Searching the Stars: The Story of Caroline Herschel* (Stroud: History Press, 2011).

30. Hoskin, *William and Caroline Herschel*, 34.

31. See R. Po-Chia Hsia, *A Jesuit in the Forbidden City: Matteo Ricci, 1552–1610* (Oxford: Oxford University Press, 2010), chapter 5, and Jonathan D. Spence, *The Memory Palace of Matteo Ricci* (New York: Viking, 1984), 180–181.

32. Catherine Pagani, *"Eastern Magnificence & European Ingenuity": Clocks of Late Imperial China* (Ann Arbor: University of Michigan Press, 2004), 62, and generally chapter 2.

33. David S. Landes points out that the Kangxi emperor and the Qing in general were never interested in horological technology, but in the "adaptation of an alien device" for their own purposes; *Revolution in Time: Clocks and the Making of the Modern World* (Cambridge, MA: Harvard University Press, 1983), 44.

34. Pagani, *Eastern Magnificence*, 62–67.

35. Sir George Leonard Staunton, *An Authentic Account of an Embassy from the King of Great Britain to the Emperor of China* (London, 1796–97), 1:491.

36. Peyrefitte, *Immobile Empire*, 73.

37. William Fraser (1720–1815) had premises at 3 New Bond Street, London, and a business that flourished between 1777 and 1818 (from 1799 in partnership with his son Alexander). The business was subsequently acquired by, successively, William Hawkes Grice and Charles Wastell Dixey, two ex-apprentices. The New Bond Street address was destroyed in the Second World War; the business relocated and continues to this day as C. W. Dixey & Son. See A. D. Andrews, "Cyclopaedia of Telescope Makers," *Irish Astronomical Journal* 20, no. 3 (March 1992), 176; and Webster's Dictionary of Instrument Makers: http://historydb.adlerplanetarium.org/signatures/search.pl?signature

=fraser&limit=100&searchfields=Signature&searchfields
=Maker+Info&searchfields=Instruments&search=1&offset=0.

38. See Henry C. King, in collaboration with John R. Millburn, *Geared to the Stars: The Evolution of Planetariums, Orreries and Astronomical Clocks* (Toronto: University of Toronto Press, 1978), chapter 12 (195–212). In 1764, Martin supplied Harvard College with two orreries, a planetarium with tullurian and lunarium attachments, and a specially commissioned grand orrery (King, *Geared to the Stars*, 200). See also John R. Millburn, "Benjamin Martin and the Development of the Orrery," *British Journal for the History of Science* 6, no. 4 (December 1973), 378–399.

39. See Laurence Worms, "Senex, John (*bap.* 1678, *d.* 1740)," *Oxford Dictionary of National Biography* (Oxford: Oxford University Press, 2004), online edition, September 23, 2004, https://www.oxforddnb .com/view/10.1093/ref:odnb/9780198614128.001.0001/odnb -9780198614128-e-25085. For Senex's *A New Map of India & China: From the Latest Observations* (1721), see https://www.raremaps.com /gallery/detail/54181/a-new-map-of-india-china-from-the-latest -observations-senex.

40. Benjamin Martin, *The Description and Use of an Orrery of a New Construction* (London, 1771), 11; and *Philosophia Britannica* (London, 1747), 370.

41. Jan Golinski, "Astronomy," in *The Blackwell Companion to the Enlightenment*, ed. John Yolton, Roy Porter, Pat Rogers, and Barbara Stafford (Cambridge, MA: Blackwell, 1991), 44.

42. Benedict Nicolson, *Joseph Wright of Derby: Painter of Light* (London: Routledge and Kegan Paul, 1968), 1:112.

43. David Solkin, "ReWriting Shaftesbury: The Air Pump and the Limits of Commercial Humanism," in *Painting and the Politics of Culture: New Essays on British Art, 1700–1850*, ed. John Barrell (Oxford: Oxford University Press, 1992), 76–80.

44. Martin, *Philosophia Britannica*, 357.

45. Jesse Molesworth, "Time and the Cosmos: The Orrery in the Eighteenth-Century Imagination," 9–10, unpublished paper given to the Early Modern Studies Group at the University of Chicago in 2016. I am grateful to Dr. Molesworth for permission to learn and to quote from his paper.

46. The identification of geography with royal power was not unique to Macartney. In the Dedication to George Adams's *A Treatise Describing and Explaining the Construction and Use of New Celestial and Terrestrial Globes* (London, 1766), Samuel Johnson wrote: "Geography is in a peculiar manner the science of princes. When a private student resolves the terraqueous globe, he beholds a succession of countries in which he has no more interest than in the imaginary regions of Jupiter and Saturn. But Your Majesty must contemplate the scientifick picture with other sentiments, and consider, as oceans and continents are rolling before you, how large a part of mankind is now waiting on your determinations, and may receive benefits or suffer evils, as your influence is extended or withdrawn" (Johnson, *Works of Samuel Johnson*, 20:521).

8

Extreme Networking

Maria Graham's Mountaintop, Underground, Intercontinental, and Otherwise Multidimensional Connections

KEVIN L. COPE

Audiences enjoy looking up. Viewers of Rodgers and Hammerstein's *The Sound of Music* acclaim its elevation-enhanced aria "Climb Every Mountain"; cerebral readers of baroque poet John Donne applaud Donne's placement of "truth" "On a huge hill, / Cragged and steep";[1] fans of television documentaries concerning extraterrestrial life imagine that spacefaring aliens need to pare a few thousand feet from their parsecs-long voyages by landing on peaks rather than in valleys. That heaps of granite can hold the attention of such diverse observers tells us much about networking in the modern age, when ease of travel and access to assorted viewpoints encourage the quick assembly of interest groups. In the thin air of international mountain tourism, along the delicate filaments of emergent long-eighteenth-century global communication systems, we detect the nowadays faint signals from the nearly forgotten Maria Graham.

If networking involves otherness and motility—if being a node on a network involves extending sensors, tentacles, or invitations toward something else, something that is *not* oneself and something that is someplace else—then Maria Graham qualifies as a premier connectivity candidate. Graham was something of a network unto herself. She lived through three more or less equally enduring identities: Maria Dundas, the curious and adventurous young woman who accompanied her swashbuckling father to a new, only slightly tamer job in India; Maria Graham, wife of a naval officer and diplomat, affiliate of the worldwide British empire, and anthropologically attuned author of travel and art history books; and Lady Callcott, wife of landscape painter and Royal Academy member Augustus Wall Callcott and writer of children's books. To these assorted identities we could also add occasional roles such as internationally acclaimed botanist (many of her collected plant material samples survive to this day);[2] illustrator (Graham executed the sketches that became the engravings in her books); geologist (her observations concerning earthquakes in Chile set off an academic feud but, in the long run, have proved to be more or less correct);[3] and even imposter (in cahoots with her publisher, she cobbled together, in an armchair and from documents, a seemingly first-hand account of a voyage to Hawaii).[4] With regard to historical periodicity, Graham also vacillates. Her emergence from a family whose keel lay deep in British naval tradition draws her back into the eighteenth century, as does her habit of citing authors such as Milton, Cowper, and Crabbe, and as does her aloof but wry style of travel writing. On the other hand, Graham's often affected, occasionally real independence (as during her period as a lone widow in South America) seems to move her forward in social history, as does her interest in Napoleon, her enthusiasm for colonial independence movements, and her taste for progressive poets such as Lord Byron. In the slender critical literature about Graham, she usually appears as a candidate feminist, although an honest survey of her travel accounts will reveal more satire of than concern for women.[5] The prolific Graham achieved, if not bestseller status, then considerable popularity; today, lost between official "eighteenth-century studies" and "nineteenth-century studies," she has disappeared from view.

During the middle period of her life—as Maria Graham rather than Maria Dundas or Lady Callcott—Graham shared with modern networks the property of apparently being everywhere in general yet nowhere in particular. Enjoying a substantial readership in England, she stood at the center of British literary culture. Yet she routinely traversed the globe, living and traveling in India, Italy, Brazil, and Chile while also occasionally happening along waypoints such as coastal Africa, Tenerife, the Juan Fernández Islands, and the Falklands. Graham spent extended but not enormous intervals in these venues—from a few days to something short of three years—yet produced voluminous, detailed accounts that give the impression of protracted residencies and cultural fluency. Although her stays were short, Graham managed to find, create, and portray extensive social networks, presenting herself as the focal point of large social arrays that, in pre-Facebook times, would normally take years to form. Networked in India, networked in Brazil, networked in Chile, and even networked in remote atolls, Graham operates at several scales: within local networks; across national and continental networks; throughout the oceanic mercantile and military network; and within the British publishing network.

All this would be impressive enough, but three more aspects of Maria Graham's worldwide networking distinguish it from the usual lists of friends, relations, and contacts developed and chronicled by other travelers. First is the role played by terrain elevation: by high mountains and deep caves. Wherever Graham goes, there are hills, mountains, mesas, spires, promontories, and assorted tall objects. Second is the use of networks and networking as literary and artistic techniques. The realities of maritime shipping or the details of diplomatic negotiations or the distribution procedures for intellectual property interest her less than the prose and painting of networking—than its literary and artistic conventions. Third is the key role of the self, of Graham, as an epitome and embodiment of her assorted networks: as a principal point of reference within large social, economic, cultural, and communications grids or even as a personification of a whole network.

The characterization and the interaction of these exceptional features—the ways in which mountains, caves, artistic technique, and self-absorption play against, balance, and counterpoint one another in an imaginative reshaping of a networked life—are the topics of this essay.

The Networked Identity: One Author, Many Faces and Places

Modern ideas about networking emphasize identity. Whether invoking server names or selecting the identifiers preceding the "@" or monitoring the stations along freight delivery lines, networks string together identifiable persons, points, or places. At the center of Maria Graham's multiplexed identity is neither Maria Dundas nor Lady Callcott, but Maria Graham, the middle interval of a segmented life involving travel to and journal accounts of vastly dispersed cultures and characters. Graham, however, is not easily pinned down behind that "@" symbol or behind any other late-Augustan predecessor to a hashtag. For one, both her writings and her illustrations look multiauthorial. She follows the usual practice of Enlightenment travel writers by inserting transcribed documents, excerpts from other authors, news stories, documents, graphs, maps, charts, and occasional literary quotations, all by way of suggesting the objectivity of viewpoint that comes from collaboration. As she sails out of All Saints' Bay in Bahia, Brazil, for example, she wonders about the location of Robinson Crusoe's plantation, suggesting that Daniel Defoe's fictional rendering overlays her factual account of the region; while in Chile, the coarse citizenry conjures up comparisons to rough characters in the novels of Frances Burney and Jane Austen, blending Graham's harsh personal judgment with softer echoes of mainstream literary figures.[6]

For two, Graham enhances her work with a bevy of images. Unlike the pictures in most travel books, which at best derive from third-party artists who have traveled similar routes and at worst emerge from hack illustrators' fantasies, most of the engravings in Graham's volumes derive from her own sketches. Graham, for example, calls attention to her authorship of the images of the figures in Elephanta Caves, Bombay (Mumbai), comparing them favorably to illustrations by previous visitors.[7] Complicating the intellectual property issue is Graham's inclusion of differing grades and qualities of illustration. Her journals include graphics derived from her own sketches but finished, refined, and upgraded by a professional engraver; "vignettes," small illustrations that appear, like printer's ornaments, at chapter or topic divisions by way of illuminating some small passage in Graham's writings; and Graham's own drawings, as directly copied to and printed from plates.[8] The various grades of illustration evidence different admixtures of Graham's hand, but, in all cases, her work has been channeled through other artisans. Thus, in the journals, there are two personae at play—Maria Graham the writer and Maria Graham the illustrator; thus, there are at least two artists—Maria Graham the sketcher and a professional engraver—at play within her illustrations; thus, Graham appears as a two-by-two, four-possibility author-artist-identity matrix.

Maria Graham's understanding of herself as a composite writer and artificer—as someone who combines sources and genres in an otherwise emphatically personal diary and who expects other artists with greater skill to retouch her sketches—informs her descriptions of wonderful phenomena in India and elsewhere. Both her subject matter and technique can be described as "networked." Seldom does Graham represent single or static phenomena, as might Joshua Reynolds when painting a portrait or Vermeer when showing a momentary situation or J.M.W. Turner when capturing the best of a sunset. Graham deploys the proverbial "cast of thousands." Her representations emerge from complicated processes, multiple actors, and plural perspectives. In Calcutta, perambulating Graham encounters thousands of luminous insects as she views "trees on the esplanade so covered with the fire-fly, as to appear like pyramids of light."[9] This scene-painting by cooperative approach—this encountering of a phenomenon generated by mass action while en route to somewhere and while seeing something else (a pyramid) in a whirl of incremental action—renders the world as arrays of distinct but cooperating, loosely affiliated beings. Graham seldom goes for a solo sightseeing outing. Setting out from Bombay, she describes a "company" comprising "one lady, two gentlemen, and three children" along with "near two hundred" attendants carrying an astounding array of equipment, supplies, and clothing.[10] Creating what art historians call a "history painting," Graham presents her motion through India as an enterprise sprawling across a logistics system that ends up seeming picturesque.

The techniques at play in Graham's "firefly" and "traveling company" passages coalesce in her urban landscapes (or seascapes). In Trincomale (now Trincomalee, Sri Lanka), incoming British naval forces create a watercolor scene:

> Yesterday we had a grand spectacle; every ship in the bay
> (among which were two seventy-fours and four frigates,) fired
> two broadsides. I never saw any thing so beautiful as the effect
> of the clouds of smoke, as they first obscured the whole horizon,
> and then gradually rolling off, left the ships reflected in the
> water, which was clear and smooth as a mirror. Nor were the
> thundering reverberations from the rocks less striking, amidst
> the grand silence and calmness of the nature around.[11]

This truly multimedia event brings together smoke, sea, sky, sound, munitions, and sensibility in a massive convergence. Unlike Romantic practitioners of synesthesia, Graham preserves the distinctness of the elements in her images. Impressions emerge from a network of events in which the separate components—the nodes—retain their independence and separation. The blare of cannons may seem incompatible with the enjoyment of fluid smoke and wafting water, but network-emergent events are all about the connecting of disparate participants. In Barrackpore, near Calcutta, Graham unveils an even more complicated scene in which fireworks play against funeral pyres along the shore while pagodas and exotic trees plunge into and soar above the water and while white moonlight dances on the flickering river and competes with dazzling white buildings.[12] To produce this highly motile as well as conceptually heterogeneous scene, cultures, buildings, peoples, watersheds, energy sources, and even planetary satellites must converge into a visual ensemble.

Long Views, High Hills, Fitting Frames: Connecting Orient and Occident from Afar

Rather than merely reporting facts, Maria Graham unveils "factual landscapes." Her composite literary-pictorial art opens views into framed panoramas. In these wraparound views, multitudinous persons and processes converge in an informative impression that includes facts, figures, and charts. A sort of Wordsworth with an abacus, Graham surveys broad sweeps of human experience while also recording the net profit of shipping voyages or monitoring the gross national product of exotic nations.

This concern for points of reference points up a major concern of this networking orientalist: *points*. The physical net that supports the networking metaphor is an array of intersection points linked by filamentary connections.[13] Most of what we see as a "net" is empty space that, in turn, is outlined by point-dependent links. Maria Graham extends the point-and-space motif to geological proportions. Fundamental to Maria Graham's geography are the most extended of all points: mountains, hills, promontories, mesas, bluffs, spires, and outcrops of tall rock. The hard verticality of Graham's world view is intensified by her attraction to all tall objects, whether high buildings, soaring trees, towers, or even long-legged wading birds.

Graham, an ardent apostle of the protuberant, finds, explores, and adores elevated land masses wherever she goes. Her India journal opens with an engraving of the Temple of Maha Deo in Bombay (Figure 8.1), a composition in which a tall temple shoots straight out of the flat ground alongside two tall topiary trees that guide the eye to clouds floating overhead.[14] Graham's preface (vii–viii) tags the carved mountain caves at Elephanta, Salsette, and Carli as India's destinations of highest interest. Within her first seven pages, Graham describes tall stones used for beating laundry; a

FIGURE 8.1. M[aria]. G[raham]., "Temple of Maha Deo in Bombay," etched by James Storer, from Maria Graham, *Journal of a Residence in India*, 2nd ed. (1813). Photograph by the author.

mighty banyan tree, the illustration of which includes a mountainous background (Figure 8.2); and Sion Fort, a garrison "on the top of a small conical hill" overlooking mountainous Salsette island.[15] Protrusions poke into nearly every page of Graham's journal. Readers discover a colossal stone figure of Cotta Rajah; view a high, projecting stone bluff surmounted by a Muslim saint's tomb (Figure 8.3); admire the towering bloom of the already tall talipot palm; appreciate the cinnamon, a shrub tall enough to pass as a tree; and witness the illumination of palm trees by torch-bearing runners: "The effect of this illumination surpassed that of any I ever saw. Sometimes the straight tall trunks of the palm-trees, whose fern-like heads remained in shadow, seemed to represent a magnificent colonnade; sometimes . . . they appeared like some enchanted bower."[16] One of the few people to have strong preferences with regard to the layout of harbors, Graham applauds the docks at Bombay owing to their adaptation to high tides.[17]

Maria Graham's mountain mania includes an intense interest in open space. A painter like Salvator Rosa may carefully portray

FIGURE 8.2. M[aria]. G[raham]., "Banian Tree," etched by James Storer, from Maria Graham, *Journal of a Residence in India*, 2nd ed. (1813). Photograph by the author.

FIGURE 8.3. M[aria]. G[raham]., "Temporary Bridge & Bungalo at Barbareen" (tomb of a Muslim saint), etched by James Storer, from Maria Graham, *Journal of a Residence in India*, 2nd ed. (1813). Photograph by the author.

peasants in valleys or may take a midrange viewpoint halfway up a hill, but Graham views elevated scenes across long, open distances. While cruising the Malabar coast, Graham looks at coastal elevations from a position far out on the high seas:

> Cape Comorin, and the islands in its neighbourhood, make, from sea, like a high rocky point, and from thence the mountains rise as we advance towards the north. In some places they are so near the shore, that they literally seem to overhang it; in others they recede a few miles, leaving space for towns, villages, and fields. They are almost clothed to the top with "majestic woods of vigorous green;" and it is only here and there that a wide tract of jungle grass, or a projecting rock, interrupts the deep hue of these ancient forests. At the foot of the ghauts, the white churches of the Christians of St. John's and of the Portuguese, appear now and then among the coco-nut woods which fringe the coast, and mix agreeably with the fishermen's huts, the native pagodas, and the ruined forts of decayed European settlements.[18]

In this quintessential passage, Graham appreciates the mountain range from afar while seemingly eliminating distance. She offers a surprising amount of detail for one so far offshore and enjoys the illusion of propinquity created by the overhanging mountains. Space, points, and connections come together as hubs of human activity ("towns, villages, and fields") dot and define the spacious landscape. "Projecting rock[s]" enhance the celebration of protuberance. Primitive but contemporary fishermen's huts complement decaying remnants from advanced western civilizations; processes of intercultural and ecumenical exchange connect not only the ethnically diverse settlers and natives, but also points in time.

Similarly, distant but detailed mountain scenes occur again and again in Graham's journals. Sailing away from India and along the Isle de France (Mauritius), Graham applauds volcanos, stones, and caves while delving into ethnic relations, mingling of races, and international commerce;[19] cruising along the coast of Africa, she hails a "high and mountainous" continent;[20] passing Cape Town, she relishes the long view of Table Mountain while reviewing the intricate relations between the English, Dutch, Hottentots, French, and Moravians;[21] visiting Rome, Graham reverses her perspective, looking down from the ruins of Sant'Angelo on a misty prospect of the eternal city while quoting John Milton;[22] touring the Roman hills, Graham intersperses elevated views with allusions to authors and painters such as Ann Radcliffe, Virgil, and, again, Salvator Rosa;[23] arriving in Chile, she exults in the longest of views, exhaling that she "can conceive nothing more glorious than the sight of the Andes" while applauding sixty-foot palm trees.[24] In Chile, Graham takes in sublime scenes at nearly hundred-mile distances even while celebrating their apparent proximity:

> The high mountains of Switzerland are always seen from a point extremely elevated; but here, from the sea-shore, the whole mass of the cordillera rises at once, at only ninety miles' distance. This gives a peculiarity to the landscape of Chile which distinguishes it, even more than its warm colour, from any I have seen.[25]

This and a second long-distance encounter with the Andes[26] occur between the intimate, immediate space of emotional responses and the long views from distant ridges, all while Graham notices local cultural phenomena such as foods served by local chefs.[27]

Juxtaposing of local detail against mountain panorama is more than mere happenstance. Zooming in and zooming out occurs habitually. Within the space of three paragraphs, for example, Graham zooms from an impression of Bahia, Brazil, as "magnificent in appearance from the sea" to a close-up shot of a dystopia choked by narrow streets, "bearers, dogs, pigs, and poultry, without partition or distinction," and detritus thrown from every window.[28] "Zooming" may not be an adequate term for Maria Graham's descriptive technique, for Graham eliminates the time lapse that goes with the cinematic zoom, presenting near and far

FIGURE 8.4. "Larangeiros," drawn by Maria Graham, engraved by Edward Finden, from Maria Graham, *Journal of a Voyage to Brazil* (1824). Courtesy of the Special Collections Department, University of Aberdeen Library.

simultaneously and synonymously. The illustrations in Graham's South American journals, for example, present distant, background mountains as central points of interest: as both far-off objects of admiration and as the immediate, principal topic of an illustration. Her illustration of the citrus-lined valley of Larangeiros shows the river, the road, a traveler, and assorted fauna at close range and in sharp focus, but it also shows, in the center of the panel, a high, rocky dome in hazy light and in soft focus, at once suggesting its remoteness but also making it the most assertive, attention-gathering, and, in sum, *present* element in the picture (Figure 8.4).[29] This technique is reiterated in *The View of the Corcovado*,[30] where detailed fronds bring the peripheral palms into seemingly immediate focus while the blurrier, distant Corcovado spire grabs the eye and shoots into the foreground (Figure 8.5); and again in the rendering of San Cristovaõ, where a remote mountain range in dappled light overtakes the curious interplay of indigenous

FIGURE 8.5. "View of the Corcovado," drawn by Maria Graham, engraved by Edward Finden, from Maria Graham, *Journal of a Voyage to Brazil* (1824). Courtesy of the Special Collections Department, University of Aberdeen Library.

FIGURE 8.6. "San Cristovaõ," drawn by Maria Graham, engraved by Edward Finden, from Maria Graham, *Journal of a Voyage to Brazil* (1824). Courtesy of the Special Collections Department, University of Aberdeen Library.

person and abbey in the foreground (Figure 8.6).[31] In still other mixed anthropological-geographical illustrations such as the view of *Rio, from the Gloria Hill*,[32] Graham portrays urban development and places tokens of commerce such as ships in the harbor but also positions mountain ranges across and above the horizon, making them equally present with the local subject matter of her illustrations (Figure 8.7). This conflation of the distant with the local and this ocular bypassing of huge blocks of space (and travel time) befits Graham's "networking" mentality: her love of linking the near with the far, of celebrating long-range relationships, and of incorporating open space and time in a way that makes it less daunting. If we forget about electricity, devices, and the Internet and instead focus on her method, Graham's illustrations seem not all that different from the World Wide Web, which celebrates its immensity at the same time that it presents simulated experience in a way that dismisses the bigness of the nominal *world*.

FIGURE 8.7. "Rio, from the Gloria Hill," drawn by Maria Graham, engraved by Edward Finden, from Maria Graham, *Journal of a Voyage to Brazil* (1824). Courtesy of the Special Collections Department, University of Aberdeen Library.

The Straight and Narrow and Never Disoriented: Maria Graham's Wiry World

Maria Graham characterizes one of her Indian destinations, Compowli, as lying amidst "amphitheatres of hills."[33] Whether in words or pictures, Graham presents all sorts of experience, from diplomatic exchanges to earthquakes, as acts on a mountain-surrounded stage. The moment she lands in Tenerife, for example, she frames herself in such a setting. She grabs a saddle, mounts a mule, and immediately ascends the mountain, delivering herself to an elevated, stony amphitheater.[34] Graham stands out among travelers by standing down and staying in: by conducting her journeys in straitened, theatrically sized spaces.

Despite her easy outdoorsiness—her habit of casually scaling mountains on her way to afternoon tea—Graham thrives in narrow spaces. This most filamentary of authors is always slipping through tight passages or scrambling along narrow trails. One of her first stops in Bombay is a "cleft in a rock, so narrow, that one would wonder how a child could get through it"[35] through which throngs of pilgrims pass by way of seeking atonement; at Compowli, she marvels at a nearly perpendicular ghaut;[36] in a comical as well as courageous moment, she straps into a narrow saddle atop the mountain-like back of a camel;[37] in Chile, Graham scours the hill trails behind her house, marvels at the narrow aqueduct channels etched into the rocks by the caciques, and treks the tight Mount San Miguel trail.[38] Travel is less about taking the path of least resistance or the safest conveyance than about thinness: about following the narrow social, geological, cultural, and artistic wires that link peoples and societies.

The geological traces of Graham's search for the slender are easy to find. Graham has a way of staging her scientific fieldwork in nooks, crannies, cracks, and crevices. Following the great Chilean earthquake of November 1822, Graham makes a combined social, diplomatic, geological, and relief excursion with Lord Thomas Cochrane, the picaresque Anglo-Chilean admiral, part-time virtuoso, and swindler, to view the aftermath of the shock:[39]

> In the evening I had a pleasant walk to the beach with Lord Cochrane; we went chiefly for the purpose of tracing the effects of the earthquake along the rocks. At Valparaiso, the beach is raised about three feet, and some rocks are exposed, which allows the fishermen to collect the clam, or scollop shell-fish, which were not supposed to exist there before. We traced considerable cracks in the earth all the way between the house and the beach, about a mile, and the rocks have many evidently recent rents in the same direction: it seemed as if we were admitted to the secrets of nature's laboratory.[40]

It is altogether characteristic of Maria Graham to fancy that her evening stroll should include an entrance into "nature's laboratory" as well as to revel in an earthquake-induced opportunity to hobnob with the influential while attending to cracks, lines, rents, and slits in the earth. Graham inventories precious minerals and gems within these narrow breaches while networking with the Chilean military top brass. Earlier during her Chilean stay, she followed a trail behind her own home and, "skirting the hill for about a furlong," she discovered a ravine full of "great blocks of granite" through which a narrow rivulet of water splashes its precarious way.[41] The recurring interplay of projecting granite blocks, narrow paths, bouncing water, geological speculation, and diplomatic activity pounds out a staccato rhythm—a verbal, even sensual reiteration of the network motif, with all its hard knots, quick paths, jolts of energy, and lucid intervals.

Another dimension of Maria Graham's narrow paths is just that: another dimension. Columbus may have crossed the flat ocean and circumnavigating Magellan may have gone from point A to point B and then around again to point A, but Graham also ascends and descends. In Bahia, which Graham pans as the worst town in Brazil, the social, cultural, physical, and even sanitary outlook improves "as we ascended from the street" and climbed to high ground, where the consul lodged.[42] Given Graham's verve for verticality, it is no surprise that she locates the sine qua non of poor Bahia, the "handsome" opera house, "on the highest part of the city, and the platform before it commands the finest view imaginable."[43] Stepping up the narrow passages of Bahia's blighted streets, Graham ascends to her favorite position on an elevated as well as space-limited platform, there to admire the actions and relations below her while keeping them artfully absent.

Supernodes: The Author as Network

An old idiom talks about being "caught between a rock and a hard place," but stone-seeking Maria Graham is seldom caught anywhere, even in those narrow rocky passages that she loves to traverse. Her favorite habitats—ledges, trails, mountains, and long-distance views—all involve motion. Roads provide the platforms on which coaches roll; panoramas unfold in far-off venues at the end of journeys. As the image of a net provides a stable icon for the energetic process of connecting to distant people and places, so Graham's anecdotes and sketches fix an event in time while moving toward the next topic, referent, or event.

Graham's motility is easily observed in her South American journals, where she floats effortlessly through the most diverse and occasionally warring groups, factions, and cliques. Wherever she is, whatever she does, she is also on her way to the next venue. One minute, she parties with the patriot rebels; the next moment, she socializes with the Portuguese Prince Royal;[44] always, she begins moving to the next milieu. Graham seems marvelously at home in mixed, slightly awkward English-Portuguese soirées, where she quietly smiles at the unpolished colonial gentry while thinking about the cleverer people back home.[45] Stuck but not altogether unhappy among the strange characters who routinely travel whole hemispheres yet who linger in provincial environments, Graham merges conceptual motion and aesthetic stasis by looking and thinking every which way while focusing on local dramas.

Bifocal and multiplexed perspectives—seeing the world from multiple viewpoints while creating a comprehensive, wraparound view of an event, culture, person, or phenomenon—dominate Graham's writings. Even as a young woman, Graham could see the culpability of both parties to the colonizing process while also assessing the merits of colonialism as a whole:

> If I could be assured that the communication with Europe would in ever so remote a period free the natives of India from their moral and religious degradation, I could even be almost reconciled to the methods by which the Europeans acquired possession of the country.[46]

So concise a statement stands as an enduring judgment while ranging back and forth across two conflicting but collaborating cultures and while traversing the vast physical and conceptual distances between two worlds. It zooms around multiple perspectives to wrap up the Euro-Indian saga in a compact utterance. Graham is the unrivaled master of sudden intercultural condensation. She quips that a pundit told her that philosopher George Berkeley "must have been a Vedanti Bramin [sic] in his pre-existent state,"[47] thus slamming together Berkeley's offbeat immaterialism with colorful India and its esoteric theologies in a one-off witticism that she attributes to a third party. In a more extended account of a visit to an Islamic house,[48] Graham builds up a bicultural, remarkably fair view of cultural differences by recounting how Muslim women responded to and interpreted her Englishness. A similar technique operates in the portrait of "Mrs. A.," Graham's hostess in Salsette, a French woman who inherited an estate on that urbanized island. Despite deeming her a friend, Graham describes Mrs. A. in a coolly sympathetic, detached third-person voice, sedimenting her portrait between layers of observed good deeds and discretely opining that there should be "a few more such European women in the East, to redeem the character of our country-women, and to shew what English Christian women are."[49] Indian and British women receive both subtle judgment and due recognition in a condensed verdict.[50] That judgment caps a friendly but detached account in which the "friend" speaks only once, to articulate another author's aphorism not in English, but in French. In the long interpolated tale of Sevajee, legendary scion of the Mahratta dynasty, Graham assumes her favorite elevated, bifocal viewpoint on both Indian history and Western morals, creating the illusion that she orbits around a whirlwind of public affairs that has been spinning since dynastic times. Graham amplifies this mixed sense of engagement and detachment by running the Sevajee story up to the end of Sevajee's life in 1680, then dribbling out an assortment of postnarrative details so as to extend the sequel for this yarn to the present time, where it lightly touches on Graham's own life.[51]

Visualizations of Internet usage show that, in any network, some nodes outweigh others with respect to usage and connectedness (Figure 8.8). The world that Maria Graham describes is unequally weighted with regard to informational potential. Graham builds her picture of oriental and other outlandish places not by evenly panning across everything but by quick presentations of

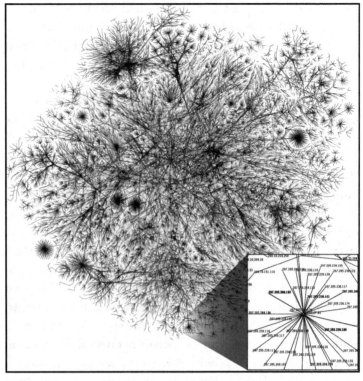

FIGURE 8.8. Visualization of Internet Usage. From Wikipedia (Creative Commons Attribution Generic License 2.5).

experiences laden with implications, insinuations, associations, and connections. For example, she captures the colonial mentality by portraying one incident, in Caliane near Bhandoop (Bhandup), in which one naïve English traveler suggests that some learned authors conjecture an ancient Greek influence on the area and proposes searching for antiquities. This misguided quest the travelers finally "gave up" as "fruitless" despite having "determined to see antique Greece at every turn" while battling intense heat.[52] Graham captures the culturally composite experience of Brazil in an anecdote and engraving of one Doña Maria de Jesus, a cross-dressing woman who masqueraded as a man so as to join the army, who sports a hybrid military costume comprising a Portuguese imperial uniform and a highland kilt, who won the order of the cross, and who provides Graham with an occasion to satirize Scotland (Figure 8.9).[53] The tale of Doña Maria comes as a simple one-two punch—or, as we might say today, with two clicks—comprising a paragraph-length anecdote and an unforgettable image. Doña Maria emerges as a "supernode," an especially rich condensation of viewpoints and possibilities. Around such supernodes, Graham weaves her rendition of the colonial world.

The one party linked to everything mentioned in a book is its author. With their love of inventories—who could forget the manifests that Robinson Crusoe assembles as he salvages goods from his wrecked ship?—long-eighteenth-century narrators easily slipped into the omniscient narrator role, monitoring an assortment of plots and subplots from an all-seeing position. For networking Maria Graham, an author is both more and less than omniscient: both more immediate than a distant panoptical observer and less domineering than a quasi-divine story manager. In her own semidetached way, Graham presents herself as part and parcel of the network of colonial experience, as one of those supernodes like Doña Maria de Jesus, who both tell and epitomize stories, who embody arrays of experiences and assortments of perspectives. Early in her stay in Bombay, Graham shows herself reconnoitering the Bombay docks. She speaks knowingly of the full range of international trade, of goods such as horses, pearls, coffee, gums, honey, ghee, dried fruits, attar of roses, tobacco, rosewater, Schiraz wine, books, slippers, shawls, cotton, wheat, rice, cattle, pottery, gems, coconuts, timber, pharmaceuticals, hardware, and assorted Chinese knickknacks, all carried by shippers from around the world. By reducing herself to one more object in this smorgasbord of world commerce and by placing herself in the middle of a flurry of activity, Graham portrays herself as a kind of living node: as a personal index to everything that is going on as it passes through her oriental world. She herself becomes the most visible point on the trade network.

Graham's self-fashioning as a living network explains the avatar-like quality of the personal presence that she projects into

FIGURE 8.9. "Doña Maria de Jesus," drawn by Augustus Earl, engraved by Edward Finden, from Maria Graham, *Journal of a Voyage to Brazil* (1824). Courtesy of the Special Collections Department, University of Aberdeen Library.

her journals. Ancient lawgivers like Lycurgus and Solon elicited order wherever they went; culture heroes like Moses or George Washington possessed the personal charisma required to get results from rambunctious new nations; Graham instigates social networks wherever she appears. Within one day of alighting in Rio de Janeiro, she announces that she has "begun house-keeping on shore," seemingly instantly establishing a complex supply chain for a wide range of foods and goods.[54] Two days later, she summarizes December 20 as "spent in paying and receiving visits in the neighbourhood"[55]—in fervid networking. Graham's past experiences conveniently flood her inbox when Lord Amherst conveniently happens by on a frigate bound for India,[56] as if Graham herself were some sort of interoceanic reloading port. Envoys casually strolling the beach trot directly to Graham to deliver urgent world news such as the recognition of several Latin American states by the American government.[57] Always seeking centrality, Graham injects herself into (somewhat skeptical) local women's groups in order to learn the secrets of native pottery[58] and drinks *matee* (yerba mate) to insinuate herself with assorted social circles; always on the prowl, she links to any group whose lore she might annex into her journals.

Maria Graham would seem authoritative were it not for her habit of disappearing into the filaments and vacant spaces of her interconnected world. She likes to look and wonder at networks (or network-derived products) that stand outside her experience or that exclude worlds with which she is conversant. In Santiago, Graham scrutinizes the home of one hostess, Madame Cotapos, inventorying not only its contents but the logistic chain that delivers them. "The rooms are handsomely but plainly furnished; English cast-iron grates; Scotch carpets; some French china, and timepieces, little or nothing that looked Spanish, still less Chileno."[59] Graham has a perverse admiration for a house that epitomizes global trade but from which every last trace of native, Spanish, or Chilean identity has disappeared. The purely networked Cotapos house captures Graham's own aspiration to be so much part of a network as not to be individually noticed. An extraordinary example of this mixture of centrality and networked extension—of absorption into a global identity—is the case of Count Hogendorp, one of the many odd characters whom Graham meets. An exiled favorite of Napoleon, the picaresque Count began his career as a soldier of fortune in Frederick the Great's army. This perpetually displaced Dutchman subsequently served as governor of east Java; traveled to the English colonies in India; was swept up, during the annexation of Holland, into the service of Napoleon, who promoted him to colonel; served the diminutive emperor as governor of Poland; likewise served as Napoleon's prefect in Hamburg; and, finally, with the collapse of Napoleon's regime, decamped to Chile. If this mad multiplicity of identities were not enough, a footnote reports that, "on undressing him after death, his body was found to be tattooed like those of the natives of the eastern islands."[60] Festooned with lines and whirls of an exotic stripe, Hogendorp's body very literally bore the inscription—the trademark—of an identity both absolutely singular and completely dispersed in global interculturalism—in networking.

Human Peaks on the International Social Network

That old idiomatic description of a connection-seeking person as "having his or her antennas out" applies in tangible ways to Maria Graham. Graham likes just about anything that stands out, sticks out, stabs the sky, or otherwise reaches up. People are important to Graham, but not because she loves everyone equally. Her self-presentation as the major node in a network reaching from the Teer of Arjoon to Tierra del Fuego implies a backdrop of thousands of persons engaged in worldwide commerce and diplomacy. Her journals single out only a small number of persons—a hundred or two at most. As she uses mountain backgrounds to suggest that a venue stands out among the millions of hectares composing our planet, so Graham summons up small populations that stand out, persons whose landmark-like presence allows for navigation through her world of anecdotes.

Graham has a special genius for finding people and stories that rise out of the narrative landscape. Her rendering of a Chilean apothecary, for example, combines the caricature skills of Samuel Butler or Thomas Overbury with the pharmacy setting in Shakespeare's *Romeo and Juliet*:

> I fancy it [the apothecary's shop] must resemble an apothecary's of the fourteenth century, for it is even more antique looking than those that I have seen in Italy or France. The man has a taste for natural history; so that besides his jars of old-fashioned medicines, inscribed all over with celestial signs, oddly intermixed with packets of patent medicines from London, dried herbs, and filthy gallipots, there are fishes' heads and snakes' skins; in one corner a great condor tearing the flesh from the bones of a lamb; in another a monster sheep, having an adscititious leg growing from the skin of his forehead; and there are chickens, and cats, and parrots, altogether producing a combination of antique dust and recent filth, far exceeding any thing I ever beheld.[61]

Such peaks of perversity protrude whenever Graham connects to a new social network. Wandering among fisher-folk, she points out the case of a ship's master who was thrown into prison for completing a voyage from Lima so quickly that the superstitious Spaniards suspected diabolical assistance.[62] Similar intercultural satire arises around one Don Fausto del Hoya, a Spanish loyalist captured by the adventurous Lord Cochrane, who seems perfectly at ease living the good life of a trusted "at large" aristocratic prisoner with abundant liberties and extensive social connections,[63] and in the critique of the "sad proportion in the English society here of trash."[64] In Quintero, Graham runs across her protector Lord Cochrane's eccentric Spanish secretary:

> There is something not unpleasant in the broad Lincolnshire dialect which gives an air of originality to his thoughts, as well as his stories. He affects a singularity of dress: sometimes a loose

shirt and looser trousers, nankeen slippers, black fur cap, and a sash, form the whole of his habiliments; at other times, wide cossack trousers, a blue jacket, real gold buttons, a small pair of epaulettes, and a military cap, and the sash tight round his waist, adorn him;—rarely does he condescend to wear a neck-cloth, even when the rest of his dress is in conformity with common usage; but when in full costume, his thin pale person-age, and eye with an outward cast in it, are set off by a full suit of black, with shiny silk breeches that look like CONSTITU-TIONAL CALAMANCO (v. *Rejected Addresses*), enormous bunches of ribbon at the knees, and buckles in his shoes. . . . Then [when previously governor at Esmeraldas], his body was painted, his head adorned with feathers, and his clothing as light as that of any wild Indian.[65]

The extraordinary detail in this description not only proves the uniqueness of this character but makes him stand out. Graham selects the Spanish secretary as a focal point, then zooms in on each and every detail by way of underlining the extraordinary interplay of cultures in a character connected to Lincolnshire, Russia, China, continental European fashion, Chilean independence movements, and an assortment of indigenous peoples. Such a description shows the scalar character of Graham's rhetoric. She zooms in on sartorial details; she zooms out to larger questions of the relation of such a character to colonial South America; allowing a button to project a world of meanings, she makes the small the mountainous.

In Graham's interconnected world, individual, small items serve as the points connecting larger cultures. In one tour de force, Graham shows how a handful of Guanche beads—beads recovered from the sepulchers of the indigenous Guanche people—can link together customs and practices in South America, Egypt, Europe, and India, all while paying homage to German explorer Alexander von Humboldt.[66] The great dragon tree of Tenerife draws together a community of palms (and associated cultural practices) extending from Bombay to Tenerife to Pernambuco.[67] People and things zoom in and out of and fade into and out of one another; in a networked world, everything is connecting to and becoming part of something else. Thus, Graham offers a description of an Islamic Indian woman that is as detailed as the description of a lady-love in any medieval romance. Attention to eye color and comments on pastry garnishing weave in and out of macroscopic comparisons of the condition of women under Islam to the experience of Western female Enlightenment writers such as Graham herself.

Burrowing Description: Dark Nets Underground and Oriental Outlooks

Maria Graham's talent for finding people who stand out brings us back to her outsized geological interests. On one hand, Graham looks for the highest peaks on the grid of experience; on the other hand, Graham's is a *burrowing* description that bores down into details that reveal cultural treasures. Maria Graham's interest in mountains extends down into the underground caverns that burrow into hills and that, after a long history of many uses, allow for cultural burrowing. Graham seeks high-level venues rich in low-level detail.

During her oriental sojourn, Graham visits cave after cave, some of which still enjoy prominence on the tourist circuit. Elephanta, Carli (Karla), Sungum, Canary (Kanheri) and many other Indian venues invite Graham to explore the interstices of the earth and, with that, the interstices between the adjacent artistic, geological, scientific, commercial, and government cultures. The definitive burrowing experience is Graham's first underground visit, to the renowned Elephanta Caves, hillside chambers on an island near Bombay (Mumbai) that early Hindus ornamented with façades, pillars, and elaborate carvings of gods, goddesses, and other sacred figures (Figure 8.10). Young Maria regards the subterranean Elephanta temple as if it were an epiphany:

> We came upon it unexpectedly, and I confess that I never felt such a sensation of astonishment as when the cavern opened

FIGURE 8.10. Elephanta Caves today. From Wikimedia Commons.

upon me. At first it appeared all darkness, while on the hill above, below, and around, shrubs and flowers of the most brilliant hues were waving in the full sunshine. As I entered, my sight became gradually more distinct, and was able to consider the wonderful chamber in which I stood.[68]

Step by step, Graham proceeds deeper into what she imagines is an abyss, with each stride developing more acute vision, finding more and more compartments, and detecting more detail. Burrowing into the cave, she devotes three full pages to the description of seemingly every figure. She captures rough sketches of the carvings, carefully providing explications of their religious significance (Figure 8.11). By the time that she concludes her visit, her sight has sharpened to the point that she can compare her sketches favorably to engravings by a German engraver, can scourge the Portuguese for damaging this monument, can speculate on the culture that produced the caves, and can take a long look at modern Brahminical culture.[69]

Two weeks later, in early December 1809, Graham arrives at the next stop on her hybrid highland-lowland journey, the caves at Carli (Karla). Noting that she is at least six thousand feet above sea level, Graham enters the Carli complex with sharpened

FIGURE 8.11. Figures in the Caves at Elephanta, drawn and etched by M[aria]. Graham, from Maria Graham, *Journal of a Residence in India*, 2nd ed. (1813). Photograph by the author.

architectural vision. "When at length we looked round, we almost fancied ourselves in a Gothic cathedral. Instead of the low flat roof of the cave of Elephanta, this rises to an astonishing height, with a highly coved roof, supported by twenty-one pillars on each side, and terminating in a semicircle" (Figure 8.12).[70] Further remarking the quality of the workmanship, the great elevation, the resulting panorama, and the purity of the water, she segues from kudos for local agriculture into an intercultural analogy in which the Carli Caves look Protestant in comparison to the Catholic-style superstitions of the icon-strewn Elephanta installation:

> The difference between the cavern temples of Carli and Elephanta is striking. Here are no personifications of the deity, no separate cells for secret rites; and the religious opinions which consecrated them are no less different. The cave of Carli is a temple dedicated to the religion of the Jines [Jains], a sect whose antiquity is believed by some to be greater than that of the Braminical faith, from which their tenets are essentially different, though many of their customs agree entirely with those of the Bramins, as might be expected from natives of the same country.[71]

One could easily mistake this critique of mainstream Hinduism and this search for primitive purity for one of Martin Luther's or

FIGURE 8.12. M[aria]. G[raham]., "Interior of the Great Cave at Carli," etched by James Storer, from Maria Graham, *Journal of a Residence in India*, 2nd ed. (1813). Photograph by the author.

FIGURE 8.13. M[aria]. G[raham]., "Entrance to the Great Cave at Kenary," etched by James Storer, from Maria Graham, *Journal of a Residence in India*, 2nd ed. (1813). Photograph by the author.

John Calvin's inventories of Catholic errors. As Graham goes farther uphill and also deeper into the earth, her capacity for linking together geographically and conceptually dispersed religions and cultures increases.

Next on Maria Graham's cave itinerary are what she calls the "Sungum" caves, now usually designated the Ellora Caves, near Pune (Poona), caves which Graham appreciates but which she compares unfavorably to the grander temples at Elephanta and Carli. Nevertheless, she deploys her burrowing approach, entering "through a natural cleft in a low rock" and forming her judgment in high and narrow straits.[72] By the time that she reaches her fourth stop, the cave at Canary (Kanheri), Graham has recovered her connoisseur's detachment, applauding or depreciating elements within this elevated cave primarily for what it shares with or what is missing with respect to the Elephanta and Carli installations (Figure 8.13). She makes use of the insular and mountaintop setting of the Canary cave to point out its suitability as a cultural outlook: "The top of the mountains commands a fine prospect over woods and mountains, and arms of the sea, to the continent of India on the one hand, and to the ocean on the other."[73] From the Kanheri precipice one can see both the colonial oriental world and the vast sea that leads to England, Brazil, Chile, and the world.

During her easygoing perambulations, Maria Graham visits half a dozen other, lesser, often ruined cave temples. No matter how degraded the facility, she presents her hilltop burrowing as an occasion for interculturalism. She comments approvingly on an old legend that, in deep antiquity, artificers from "northern countries" came to central India and "began to form the rocks" into an assortment of temples, grottos, and other public and religious structures. She regards it as "remarkable that the head-dress of the gods and principal persons . . . have not the smallest likeness to any used in this part of India, but they extremely resemble those of the countries bordering upon Tartary, and those represented in the cave of Elephanta." India's high cave temples provide an archaeological network for the diffusion and rechanneling of global cultures.

Collages and Layers: Worldwide Distribution of Processed Images

Whether or not she qualifies as an archaeologist, anthropologist, or sociologist remains open to question, but no one doubts that Maria Graham is an *illustrator*. Graham's India journal amounts to a series of local illustrations most easily seen from hills and mountaintops. Her journals move fluidly and seamlessly between illustrative written reports and telling sketches and engravings; her cave vignettes turn sketched elements into continuously moving sagas, adding literal depth to her accounts. Her literary-illustrative technique could be summed up in two words: "collage" and "layer."

The consummate example of Graham's collage-and-layer technique, unfortunately far too long to quote, is her first view of the Bombay vegetable fields. Overawed by the "great beauty" of these plantings, she enumerates, in rat-a-tat rhythm in metronome-reminiscent sentences that follow an unvarying subject-verb-object order, a panoply of edibles from around the world. "All sorts of gourds and cucumbers are in great plenty, but this is early in the season for them. Several plants produce long pods, which, being cut small, are so exactly like French beans, that one cannot discover the difference"—and so on, through a veritable Noah's ark of vegetables.[74] However simple in structure, each sentence contains two layers: an opening clause identifying and characterizing the food and a follow-on clause providing supplemental agricultural, social, historical, and culinary information. As the account continues, the vegetable fields recede behind a layered account of the entire Indian pantry, from vegetables to meats and on to breadstuffs, dairy products, fish, and fowl. India itself becomes a layer in a reckoning of foodstuffs that includes the history of imported staples such as potatoes. This colossal mural of global nutrition remains a collage: not truly a single composition, but an assembly of elements that retain their character as bits and pieces of a worldwide supply chain—a network.

The collage-and-layer technique occurs in almost every incident in Graham's journals. As she climbs to the Carli Cave, she unveils a descriptive array in which she layers geological features, fauna of assorted heights, aromas, ocean views, wild animals, and pre-Darwinian hints of the struggle to survive:

> It is impossible to describe the exquisite beauty of the landscape. High mountains and bold projecting rocks, overhang deep woods of trees unknown to Europeans. Flowering shrubs of most delicious perfumes, and creeping-plants of every various hue, form natural bowers as they hang from tree to tree, and now shewing, now concealing the distant ocean, delight the eye at every step; while here and there an opening like a lawn, with herds of antelopes, makes you forget that the tiger prowls through the overhanging forest, and that the serpent lurks beneath the many-coloured bower.[75]

What sets this apart from the usual preromantic landscape blur (or from the impressionistic paintings of J.M.W. Turner) is the exquisite concern for depth, space, scale, and, in sum, layers, as well as for the inventorying that is the hallmark of the great age of Augustan virtuosi.

Graham demonstrates enormous creativity in the assembly of deep collages. Finding differences among birds of a feather, she composes a collage of tall birds in a nearby avian menagerie.[76] While in Madras, she piles up snippet-sized illuminations of boatmen, jugglers, sword swallowers, botanical gardens (with their odd plants), and just about everything under the hot Indian sun, pasting disparate bits and pieces into a picture in which all distances are in focus and where depth of field is unlimited.[77] Graham's entry into Calcutta is likewise a fragmented depth experience. Leaving her ship and boarding a schooner, she spots the mysterious, dark island of Saugor, reputedly inhabited by fanatical devotees of Kali, the goddess of destruction; advancing up the river, "the jungle grew higher and lighter, and we saw sometimes a pagoda, or a village between the trees."[78] Later, in South America, we find Graham recommending that appreciative observers patch together an impression of the skies by reading the meteorological logs of navigators; we witness a street-level collage of all the world's wares spread out among the Rio shops; and we marvel when, in a single paragraph, Graham glues together references to the Falkland Islands, the Arabian literary tradition, Malta, Naples, Pompeii, and Africa in one compact celebration of depth and distance.[79]

Edible Collages: Or, Networked Geographies with a Culinary Flair

While traveling in Central Chile, Maria Graham offers one of her most artful collage-and-layer descriptions, beginning with the Paine river, proceeding across the open vales, describing the settlement and gardens, and then sliding up the background of mountains and even into the clouds, adding a second illustration that follows this same stereoscopic travel from extreme foreground to extreme background, all with cultivated plants at the center.[80] In this collage and in the adjoining picture, the cultivated area—Maria Graham's answer to what exobiologists call the "Goldilocks" or "habitable" zone—is the most central and yet also the least noticeable, rather like a computer in a vast network that all but loses its individuality through successfully connecting with many others. No one could miss the importance of botanical, culinary, and other garden assemblages in Maria Graham's writings. Edible, fragrant, and medicinal plants pop up everywhere, introducing a still-life element and occupying a disproportionate share of the picture plane.

Why? Despite describing one locality after another, Graham is no locavore. Whether culinary, medicinal, or research-oriented, Graham's gardens always connect to some other place. Her very food is networked. The Angostura gardens, for example, abound in Asian, European, and Middle Eastern fruits and flowers. In Valparaiso, Graham notes that gathering Chilean wildflowers "made me imagine myself in an English lane."[81] In Brazil, she takes special delight in the mixed coffee and citrus plantations, where boundaries between old, Asiatic, and colonial worlds blur in contrast-intensive plantings and where she imagines that even grafted plants spring up freely from the fertile native soil.[82] In India, Graham finds a motley assortment of English expatriate virtuosi creating experimental, often medical gardens and maintaining staffs of artists to record their accomplishments for the English audience.[83] While in transit to India, Graham delights in a horseback excursion up and around Cape Town's Table Top Mountain, where, once again at high altitude, she applauds the profusion of imported (what we would now call "invasive") blooming vegetation in a country with only one indigenous plant.[84] The discovery of oriental as well as occidental plants in a transit zone like South Africa comes as no surprise, for, wherever she goes, Graham connects, and connects botanically, to Asia. In Rio de Janeiro she discovers a patch of the composite Orient:

> This garden was destined by the King for the cultivation of the oriental spices and fruits, and above all, of the tea plant, which he obtained, together with several families accustomed to its

culture, from China. Nothing can be more thriving than the
whole of the plants. The cinnamon, camphor, nutmeg, and
clove, grow as well as in their native soil. The bread-fruit
produces its fruit in perfection, and such of the oriental fruits as
have been brought here ripen as well as in India. I particularly
remarked the jumbo malacca, from India, and the longona
(*Euphoria Longona*), a dark kind of lechee from China.[85]

An accomplished specimen collector, Graham knows that there
is more to cultivating an alien species than finding immigrant
families from its home nation to work the fields. She presents this
anecdote less as a field study in agricultural science than as a
demonstration that people and products travel together and that
at least the affluent can create the captive, stationary semblance of
international economic and biological networks—that the natural,
social, and economic systems in South America can connect with
those of the Far East.

The discovery of such intensely networked patches is one reason that food collages are so important for Graham. Produced and
harvested, food is literally a cut above ordinary plants and animals.
The making of food requires processing within economic and cultural systems. While passing through Tenerife, Graham unveils a
fantastic layered collage in which plants and foodstuffs from India,
China, Japan, Europe, South America, and just about everywhere
else converge in a massive edible potpourri. "Here, the sago-palm,
platanus, and tamarind, as well as the flowers and vegetables of the
north of Europe, flourish," while "wheat, barley, a few oats, maize,
potatoes, and caravansas all freely grow," as dietary traditions from
the new world, the Orient, and Europe converge in a node where
the distinctions between aristocratic and peasant diets also seem
to rearrange themselves.[86] The Valparaiso market, likewise, yields
a kaleidoscope of vegetables that rival the cabbages of Lambeth,
the figs of the Orient, the potatoes of the indigenous Chileans, and
just about anything else that the world produces.[87] Whatever enters
Graham's mouth connotes an ensemble of possible connections
among diverse cultivation practices and world cuisines. A lowly

potato can become as high a point in her network as any of Alexander von Humboldt's volcanos.

More than Underground Cables: Buried Oriental Networks as Intercultural Utopias

Low places with high mountain backdrops; buried potatoes and soaring palms; riotously colorful marketplaces and delicate high-country flowers; cavernous temples atop promontories; harbors barely deep enough to float a ship; India, England, Italy, Brazil, Chile: what does all of this have in common or add up to, other than pertinence to the relentlessly varied life of Maria Graham?

The ever-networking Maria Graham offers something of an answer in some of her most offbeat as well as spectacular vignettes. Once again changing altitudes as she descends from Malabar Hill, she discovers not only one, but an interlocking array of multicultural burial grounds:

> [Along] the road from Malabar Hill to the Fort of Bombay . . .
> is the general burial-place of all classes of inhabitants. That of
> the English is walled in and well kept; it is filled with pretty
> monuments, mostly of chunam, and contains many an unread
> inscription, sacred to the memory of those who, to use the
> oriental style, "had scarcely entered the garden of life, much less
> had they gathered its flowers." Next to the British cemetery is
> that of the Portuguese, after which follow those of the Arme-
> nians, the Jews, and the Mahomedans, with the few Hindoos
> who bury their dead in regular succession; they are all over-
> shadowed by a thick coco-nut wood."[88]

A postmortem multicultural utopia where everyone (very) quietly gets along, the burial place is also a network. Not a single cemetery but a chain or web of separate, culturally informed, adjacent plots, the cemetery provides a still collage of Indian intercultural dynamism. Low but on a road to a high place, festooned with peaked monuments at more or less regular intervals, the burial place looks somewhat like webbing—like a network. Lest it seem that the burial place is a one-off oddity, it should be noted that Graham elsewhere discovers several such mixed subterranean and protuberant collages. In Trincomale, she sails into port comparing herself to Sinbad, lauds a "bold projecting rock" harboring a ruined Hindu temple and some caves, compares the place to a Scottish loch, and explains how a British admiral deployed a band of Chinese experts to transform a rocky ruin into one of those internationalized vegetable gardens of which Graham is fond.[89] High, low, various versions of the East, a British version of the West, and a worldwide web of expertise come together in the renovation of a rocky ruin. In another tour de force, Graham presents the Kanheri Caves as a "subterraneous city" immured within a high cave that is nevertheless within a ravine, which ravine, topped with soaring trees, leads upward to a grand view of India and the surrounding seas.[90] Graham even finds a mountaintop town burrowed into the rocks in less exotic Italy.[91]

From these mixed, seemingly catachrestical high and low spots, Graham evolves her vision of the improvement of humanity through social connectedness and through the layered, composite, networked evaluation of persons, places, things, and events. Deep in the otherwise elevated Elephanta Caves, Graham riffs on the wonders and novelties she encounters, imagining her way back to the time of their construction, when an elevated observer could apprehend all the diverse arts, crafts, trades, people, and support systems required to compile such a project. She tops it all off by connecting that golden age of Indian arts with the European Enlightenment and then finally unfurling her Euro-Indian-Enlightenment allegory by way of condemning the contemporary, oppressive government of India.[92] Going still further, she complements the Elephanta allegory with a similarly complex array of interpretations elicited from the Carli Cave, the composite double allegory being used to doubly critique the superstitious Hindus and the abusive British.[93] The two descriptions together—separated by eight pages in Graham's book and fifty miles on the ground—create a reciprocating account of India, its complexities, and the world in which the "Orient" operates. By stretching webs of meaning across enclosed but related spaces, these two accounts communicate volumes of information in a minuscule space.

Capstone: Draping Smooth Networks over Eastern and Western Peaks

As Maria Graham and her friend and counterpart in internationalism, Lord Cochrane, leave South America, Lord Cochrane makes one last climb to the top of a pinnacle on Juan Fernández Island, from which he retrieves a piece of high-flying lava apparently from an adjacent, higher cone that he wishes he could have climbed.[94] Close, medium, and long range come together as the adventurous admiral takes a stone once launched from a mountain as a souvenir of a climb to moderate heights that left him yearning for even more elevated views. Likewise, when Maria Graham leaves behind any close-up view, whether of cave art or Chilean badlands, she invariably ascends to a high, panoramic outlook that seems to put her in sight or memory of some distant culture, whether England or one of the exotic places in her logbook. Mountains and lookouts provide the sense of immense space and distance that is a definitive condition of modern experience, whether expressed in John Locke's concern for traversing the gaps between "things" and "ideas" or in the rise of modern astronomy with its ethereal immensities or in the introduction of New World potatoes into Indian and Italian cuisines. For Graham, mountains provide the high baseline as well as the visual rhythm that string together diverse cultures, people, and phenomena. It is as if her verbal and graphic illustrations suggest that, were one to climb up those mountains in the far South American background, one could see India's Elephanta Caves or Atlantic Tenerife or Europe's England or maybe even that heap of polenta marketed by an Italian merchant in Valparaiso. A worldwide frame for the collage of knowledge, rock-steady mountains and the isolated caves that they contain serve as the geological signature of the interactive long-eighteenth-century world. Points holding together a delicate, highly dispersed web of acquaintance, life stories, scientific activity, commercial adventure, and political negotiation, the high spots in the life of a writer who spent a good deal of her time voyaging at sea level suggest an openness, spaciousness, and multiplicity of viewpoints that is only now being achieved in our own networked world. With all her zeal for natural history and with all her plunges into the depths of human nature, Maria Graham was nevertheless one of the first to put "artificial" alongside "intelligence": to suggest that a travel writer could do somewhat more than observing cultures A, B, and C, that travel writing was less adept at discovering new knowledge than at sketching the filaments that hold it—loosely yet artfully—together.

Notes

1. John Donne, "Satire III," *Representative Poetry Online*, University of Toronto Libraries, https://rpo.library.utoronto.ca/poems/satire-iii.

2. On the extent of Maria Graham's specimen collecting, which also included the gathering of seeds, see Carl Thompson, "'Only the Amblyrhyncus': Maria Graham's Scientific Editing of *Voyage of HMS Blonde* (1826/27)," *Journal of Literature and Science* 8, no. 1 (2015), 50–51.

3. Graham's knowledge of geology has been underestimated. She has commonly been presented as someone who happened upon the Geological Society and then got caught up in a controversy with a prominent geologist, George Greenough, who held the "Neptunist" position that oceanic action shaped the earth, a position apparently countered by Graham's observations of the effects of earthquakes. For a detailed reconstruction of the debate and for an evaluation of Maria Graham's knowledge of geology, see Carl Thompson, "Earthquakes and Petticoats: Maria Graham, Geology, and Early Nineteenth-Century 'Polite' Science," *Journal of Victorian Culture* 17, no. 3 (2012), 329–346.

4. Opinions differ regarding Graham's reconstruction of the voyage to Hawaii of the HMS *Blonde*. Although the volume generated controversy in its time, Carl Thompson recently argued that Graham attained basic scientific competence and was also navigating her way through

differing biological classification systems, the accuracy of which remained under debate in her time. Thompson credits Graham with pioneering the editing of scientific writing. See Thompson, "Only the Amblyrhyncus," 51 and 48–65.

5. The multiplicity of Maria Graham's identities—daughter of a dominant father; wife of a colonial officer; independent woman; wife of a popular, knighted, affluent artist—can lead to subtle maneuvering when it comes to considering her attitude toward women. A fine example is Betty Haglund's "The Botanical Writings of Maria Graham," *Journal of Literature and Science* 4, no. 1 (2011), 44–58. Haglund admits that some of Graham's scientific jottings can be written off "as a schoolgirl's introduction to natural history" (45). She takes a moderate course, reducing Graham's singularity by placing her in the context of an assortment of colonial wives who collected specimens and seeds, and elevates Graham in recognition of her scholarly concern for detail, especially scientific detail (see 45–47). In Haglund's final verdict, "Graham was both typical and unusual. Her interest in botany was, at least in part, a response to the popularity of the subject within early nineteenth-century British culture, but her educational background, her intelligence, and her family and marital circumstances enabled her to explore that interest in ways that went beyond those open to most contemporary women" (53). Other scholars have taken less felicitous approaches to Maria Graham and gender issues. Lucia Guerra, for example, harshly deposits Graham among an inventory of early nineteenth-century women who have ended up in problematic positions vis-à-vis gender roles. For Guerra, Graham appropriates masculine identity, specifically the kind of persona affected by German scientist-explorer Alexander von Humboldt. See "Nociones de la otredad en los textos de viajeras del siglo XIX," *Chasqui: Revista de Literatura Latinoamericana* 45, no. 1 (May 2016), 31–32. Michelle Medeiros, in "Crossing Boundaries into a World of Scientific Discoveries: Maria Graham in Nineteenth-Century Brazil," *Studies in Travel Writing* 16, no. 3 (2012), 263–285, takes a somewhat forced but nevertheless network-oriented approach by studying how Graham connected with Brazilian Empress Leopoldina to create both a social and a scientific network and to develop a "hybrid narrative"

and a "feminized" approach to the study of the new world and its natural phenomena.

6. See Maria Graham, *Journal of a Voyage to Brazil, and Residence There, during Part of the Years 1821, 1822, 1823* (New York: Praeger, 1969), 157 (henceforth abbreviated as *Voyage to Brazil*); and Maria Graham, *Journal of a Residence in Chile During the Year 1822* (New York: Praeger, 1969), 179 (henceforth abbreviated as *Residence in Chile*).

7. See Maria Graham, *Journal of a Residence in India*, 2nd ed. (Edinburgh: Constable, 1813), 57 (henceforth abbreviated as *Residence in India*).

8. For an excellent example of Graham's own, informal drawings as transferred to plates, see her sketches of the carved figures within the grotto of Mahaballiporam (today, Mahabalipuram) in *Residence in India*, inserted between pages 158 and 159. These rough images stand in marked contrast to the sophisticated engravings that illuminate her book. It has lately come to light that the trove of Indian drawings and sketches by Graham owned by the British Library extends to more than six dozen items, five dozen more than appear (in revised, upgraded, engraved format) in her account of her travels through India. Graham's original drawings also show that engravers edited and to some extent sanitized her drawings, for example replacing, in one panel, a gathering of local men with a more exotic woman in a sari and eliminating ruined items such as crumbling statues. See Lacy Marschalk, "'What Strikes the Eye': The Forgotten India Sketches of Maria Graham," *Tulsa Studies in Women's Literature* 35, no. 2 (2016): 515–518.

9. Graham, *Residence in India*, 147.

10. Graham, *Residence in India*, 59.

11. Graham, *Residence in India*, 122.

12. Graham, *Residence in India*, 142.

13. Recent scholarship has considered the vacuity of networks: their imposition of points, filaments, and connections atop much larger intervals and spaces. One accessible example is the empty world—the vacant cities, streets, and sidewalks—glimpsed by millions of Google Earth users as they peruse pictures drawn up by search requests and as they thereby create an abstracted collage of a bodiless world. As Anna Munster observes in *An Anesthesia of Networks: Conjunctive Experience in Art and Technology* (Cambridge, MA: MIT Press, 2013), 52, "the

population" of such network-emergent atlases "chooses to reside elsewhere" in much the same way that Maria Graham experiences whatever happens to be her current setting through the lens of past voyages or through the perceptions of her widely traveled colleagues. Munster draws an explicit comparison between modern quasi-cartographic, multiuser renderings of the world and the "baroque monad," a "Leibnizian" individual identity defined by perspectives on all the other perspective-holders in the universe (see 56). She cautions, however, that the baroque predecessor of the modern Internet map searcher is oriented to the outside world. The cartographically minded networker of our time, by contrast, *invokes* a whole world, an "event horizon" that may not require a reality beyond the screen. Maria Graham, during the late days of the Enlightenment, deploys a strategy between these two extremes by evoking, invoking, and provoking a global picture that appears as a whole but that assertively refers and recurs to her multitudinous experiences in scattered locales.

14. Graham, *Residence in India*, facing p. 1.

15. Graham, *Residence in India*, 1–7.

16. Graham, *Residence in India*, 91, 93–94, 98–99, 100, 102.

17. Graham, *Residence in India*, 12.

18. Graham, *Residence in India*, 107.

19. Graham, *Residence in India*, 171–172.

20. Graham, *Residence in India*, 174.

21. Graham, *Residence in India*, 174–176.

22. See Maria Graham, *Three Months Passed in the Mountains East of Rome* (London: Longman, 1820), 73 (henceforth abbreviated as *Three Months . . . Rome*).

23. Graham, *Three Months . . . Rome*, 76–78, 89–90, 155.

24. Graham, *Residence in Chile*, 113, 136.

25. Graham, *Residence in Chile*, 185.

26. Graham, *Residence in Chile*, 196–197.

27. Sublime moments and encounters with breathtaking landscapes are not uncommon in Graham's writing. According to Jennifer Hayward, scenes in which Graham is overwhelmed by an imposing prospect point up the autobiographical character of her writing insofar as they call attention to the astounded observer. Hayward notes that Graham,

whose autobiography is segmented across several visits to assorted continents, absorbs many conventions of the Gothic narrative, whether the presentation of thrilling scenery or status as a loner cut off from friends and relations. She uses those conventions to create a recognizable narrative and set of conventions that help both her and her readers to absorb the experience of the new world. See Hayward's "'The Uncertainty of the End Keeps up the Interest': Maria Graham's *Journal of a Residence in Chile* as Life Writing," *Auto/Biography Studies* 17, no. 1 (2001), 47–49.

28. Graham, *Voyage to Brazil*, 132–133.

29. Graham, *Voyage to Brazil*, facing p. 163.

30. Graham, *Voyage to Brazil*, facing p. 220.

31. Graham, *Voyage to Brazil*, facing p. 246.

32. Graham, *Voyage to Brazil*, facing p. 169.

33. Graham, *Residence in India*, 61.

34. Graham, *Voyage to Brazil*, 83.

35. Graham, *Residence in India*, 10.

36. Graham, *Residence in India*, 62.

37. Graham, *Residence in India*, 75.

38. Graham, *Residence in Chile*, 117, 214, 267.

39. With regard to seeing Chile in the context of both a world mercantile network and a world assembled out of Maria Graham's peregrine experiences, Lord Cochrane plays a key role. A man for all nations— he had decamped to South America following financial scandals in the stock market, but nevertheless acquired naval commands in Chile, Brazil, and Greece before being restored to honor in Britain— Cochrane mirrors and amplifies Graham's own career while serving as a liaison between the various populations (and levels of refinement and sophistication) with which Graham connects. See M. Soledad Caballero, "'For the Honour of Our Country': Maria Dundas Graham and the Romance of Benign Domination," *Studies in Travel Writing* 9 (2005): 112, 121–127.

40. Graham, *Residence in Chile*, 329.

41. Graham, *Residence in Chile*, 122.

42. Graham, *Voyage to Brazil*, 134.

43. Graham, *Voyage to Brazil*, 139–140.

44. Graham, *Voyage to Brazil*, 120.
45. Graham, *Voyage to Brazil*, 142.
46. Graham, *Residence in India*, 72.
47. Graham, *Residence in India*, 74.
48. Graham, *Residence in India*, 17–19.
49. Graham, *Residence in India*, 114–115.
50. A challenging aspect of Graham's style is her interweaving of approving, engaged, and sympathetic comments with detached, cool judgments. In "Beyond Fellow-Feeling? Anglo-Indian Sympathy in the Travelogues of Eliza Fay, Maria Graham, and Fanny Park," *Textus: English Studies in Italy* 25, no. 2 (2012), 134–139, Elena Spanderi identifies this vacillation between sympathy and detachment as a fundamental feature of Graham's writing and an indication of the difficulty among authors of her time in reconciling ideas about sympathy (derived from philosophers such as Adam Smith and Francis Hutcheson) with the harsh realities not only of colonial life, but of the sometimes vulgar and violent colonists themselves. Similar conclusions are reached by Krishna Sen, "A Passage to India: India and the Imperial Gaze in Maria Graham's *Journal of a Residence in India*," *Nineteenth-Century Literature in English* 9, no.1 (2005), 233.
51. Graham, *Residence in India*, 80–82.
52. Graham, *Residence in India*, 118.
53. Graham, *Voyage to Brazil*, 292–294.
54. Graham, *Voyage to Brazil*, 89.
55. Graham, *Voyage to Brazil*, 162.
56. Graham, *Voyage to Brazil*, 246.
57. Graham, *Residence in Chile*, 177.
58. Graham, *Residence in Chile*, 141.
59. Graham, *Residence in Chile*, 206.
60. Graham, *Voyage to Brazil*, 171–172 and 172n.
61. Graham, *Residence in Chile*, 133–134.
62. Graham, *Residence in Chile*, 161.
63. Graham, *Residence in Chile*, 310.
64. Graham, *Residence in Chile*, 156.
65. Graham, *Residence in Chile*, 302–303.
66. Graham, *Voyage to Brazil*, 88–89.

67. Graham, *Voyage to Brazil*, 84–85.

68. Graham, *Residence in India*, 54.

69. Graham, *Residence in India*, 57–58.

70. Graham, *Residence in India*, 64.

71. Graham, *Residence in India*, 65.

72. Graham, *Residence in India*, 78–79.

73. Graham, *Residence in India*, 117.

74. Graham, *Residence in India*, 24–25.

75. Graham, *Residence in India*, 63.

76. Graham, *Residence in India*, 143.

77. Graham, *Residence in India*, 123–127.

78. Graham, *Residence in India*, 132.

79. Graham, *Voyage to Brazil*, 91, 189, 202–203; and *Residence in Chile*, 130.

80. Graham, *Residence in Chile*, 254–256 and facing pages. Density is an important concern in Maria Graham's illustrations. The broad expanses that she portrays range from hard and dense mountains to wispy clouds. Her panoramas and vistas are notably short of occupants, indeed are often vacant. On a few occasions, persons that she envisions—for example, the old indigenous cacique who appears to her in a reverie—dissipate into the tenuous clouds. On the role of vacancy and extenuation in Maria Graham's interpretations of social and cultural phenomena and with respect to the "gaze" that a foreigner directs to a native culture, see Angela Perez-Mejia, *A Geography of Hard Times*, trans. Dick Custer (Albany: State University of New York Press, 2004), 94–96.

81. Graham, *Residence in Chile*, 153.

82. Graham, *Voyage to Brazil*, 165.

83. Graham, *Residence in India*, 145–146.

84. Graham, *Residence in India*, 177.

85. Graham, *Voyage to Brazil*, 163.

86. Graham, *Voyage to Brazil*, 84.

87. Graham, *Residence in Chile*, 132.

88. Graham, *Residence in India*, 11.

89. Graham, *Residence in India*, 120–121.

90. Graham, *Residence in India*, 117.

91. Graham, *Three Months . . . Rome*, 70.

92. Graham, *Residence in India*, 58.

93. Graham, *Residence in India*, 65–67.

94. Graham, *Residence in Chile*, 348.

Acknowledgments

We would like to thank Bucknell University Press' two anonymous peer reviewers for their constructive criticism in helping to improve this book.

Bibliography

Art Works

Alexander, William. "Hall of Audiences at Yuen-min-Yuen." Engraved by Wilson Lowry from a drawing by William Alexander. From George Staunton, *An Authentic Account of an Embassy from the King of Great Britain to the Emperor of China*, 3 vols. (London, 1796–97). Vol. 3. Plate 22. No pagination.

Anon. *Belgian Faience Polychrome Model of a Pug*. Late eighteenth century. Faience. 8.9 cm high. Christie's, London (September 16, 2012).

Anon. *Bonbonnière*. 1760–1780. Painted enamel. 5.7 × 6.3 cm. BI644. Black Country Museums, Wolverhampton.

Anon. *Bonbonnière*. Circa 1770. Faience. 14 × 25 cm. Private collection.

Anon. *Chinese Court Lady with Dog*. Tang dynasty (618–907). Terracotta. 52 cm high. Timeline Auctions, London (January 14, 2017).

Anon. *A Chinese Exported Seated Pug Dog*. Qianlong period (1736–1795). Porcelain. 24.5 cm high. Christie's, London (sale 12291, November 9, 2016; lot 491). http://www.christies.com/lotfinder/Lot/a-chinese -export-seated-pug-dog-qianlong-6033487-details.aspx.

Anon. Cover (Bow Porcelain Factory). Part of a set listed as *Mug and Cover*. Circa 1755–1760. Soft-paste porcelain. 20.9 cm high, 12.1 cm diameter. Victoria and Albert Museum, London.

Anon. *Cup and Saucer*. Circa 1750. Porcelain. Cup 3.8 cm high, 7.9 cm diameter; saucer 12.4 cm diameter. 642&A–1903. Victoria and Albert Museum, London.

Anon. *Figure*. Tang dynasty (618–907). Unglazed earthenware. 12 × 4 × 9 cm. FE.155–1974. Victoria and Albert Museum, London.

Anon. *Figure.* Early eighteenth century. Blanc de chine porcelain. 14.3×15.55 cm. C.108–1963. Victoria and Albert Museum, London.

Anon. *Figure of a Dog.* Circa 1700–1750. Porcelain with a white glaze. 15.2×14 cm. 4823–1901. Victoria and Albert Museum, London.

Anon. *Figure of a Dog.* Eighteenth century. Porcelain painted with overglaze enamels and gilt paint. 18.5×11.3 cm. FE.11A–1978. Victoria and Albert Museum, London.

Anon. (Saint-Cloud porcelain manufactory). *Figure of a Pug-Dog.* Mid-eighteenth century. Porcelain painted in enamels. 20×23.5 cm. C.317–1909. Victoria and Albert Museum, London.

Anon. *Figure of a Seated Court Lady.* Eighth century. Earthenware with sancai glaze. 37.5×14.3×15.4 cm. Metropolitan Museum of Art, New York.

Anon. *A George III Period Chinese Export Reverse Mirror Painting.* Circa 1775. Paint on reverse of mirror. 96.5×60.5 cm. Ronald Phillips, London.

Anon. *Man and Woman.* Eighteenth century. Porcelain decorated with enamel and gilt paint. 16.5×15 cm. C.12–1951. Victoria and Albert Museum, London.

Anon. (After Petrus Schenk; Meissen porcelain manufactory). *A Meissen Plate with the "Parrot and Spaniel" Pattern.* Circa 1740. Hard-paste porcelain. 23.6 cm diameter. Bonhams, London (December 7, 2011).

Anon. (Bow porcelain factory). *Mug and Cover.* Circa 1755–1760. Soft-paste porcelain. 20.9 cm high, 12.1 cm diameter. 414.109&/A–1885. Victoria and Albert Museum, London.

Anon. *A Pair of Erfurt Pug Tureens and Covers.* Circa 1760. Manganese. 16 cm wide. Christie's, London (July 8, 2002).

Anon. *A Pair of Miniature Oval Portraits.* Eighteenth century. Oil and gold paint on reverse of mirror. 8.89×11.43 cm. Martyn Gregory Gallery, London.

Anon. *A Pair of Paintings of Chinese Women Accompanied by Pipes and Pets.* Eighteenth century. Gouache on paper. 80×59 cm. Martyn Gregory Gallery, London.

Anon. (Vincennes Porcelain Manufactory). *Pair of Pug Dogs.* Circa 1750–1752. Soft-paste porcelain on mount. 20×19 cm. Artcurial, Paris (June 20, 2006).

Anon. (Longton Hall factory). *Pair of Pugs Seated on Their Haunches with Curled Tails.* Circa 1753–1757. Soft-paste porcelain with underglaze manganese decoration. 8.9 and 9.2 cm high. Mr. and Mrs. Sigmund Katz Collection, Boston Museum of Fine Arts.

Anon. (Lowestoft porcelain factory). *Pug.* Circa 1755–1760. Porcelain with manganese coloring. 9.21 cm high. 1988.985. Boston Museum of Fine Arts, Boston.

Anon. (William Duesbury and Co.). *Pug Dog.* 1760. Soft-paste porcelain painted in enamels. 8.3 cm high. 414.209/A–1885. Victoria and Albert Museum, London.

Anon. *Pug on a Cushion.* Circa 1745–1755. White salt-glazed stoneware. 4×5 cm. C.800–1928. Fitzwilliam Museum, Cambridge.

Anon. *A Rare Pair of Chinese Export Figures of Seated Pug Dogs.* Qianlong period (1736–1795). Porcelain. 17.5 cm high. Now in private collection. Christie's, Amsterdam (sale 2875, December 13–14, 2011; lot 529). http://www.christies.com/lotfinder/Lot/a-rare-pair-of-chinese-export -figures-5524008-details.aspx.

Anon. *Woman Holding a Pekingese.* Tang dynasty (618–907). Colored terracotta. 49.2 cm high. Kyoto National Museum, Kyoto.

Borget, Auguste. *Moonlit Scene of Indian Figures and Elephants among Banyan Trees, Upper India (probably Lucknow).* Circa 1787. Oil on Indian hardwood. 23.3×41 cm. Paul Mellon Collection. Yale Center for British Art, New Haven, CT.

Boucher, François. *Audience of the Emperor of China.* 1742. Oil on canvas. 40×65 cm. Musée des Beaux-Arts et d'Archéologie, Besançon.

———. *A Young Lady Holding a Pug Dog.* Circa 1745. Oil on canvas. 34.5×28.6 cm. Art Gallery of New South Wales, Sydney.

Coeman, Jacob. *Pieter Cnoll, Cornelia van Nijenrode and Their Daughters.* 1665. Oil on canvas. 132×190.5 cm. Rijksmuseum, Amsterdam.

Collins, Richard. *The Tea Party.* Circa 1727. Oil on canvas. 100×120 cm. Goldsmiths' Hall, London.

Daniell, Thomas. *Temple, Fountain and Cave in Sezincote Park.* 1819. Oil on canvas. 101.4×120 cm. Yale Center for British Art, New Haven, CT.

[Daniell, Thomas, or William Daniell]. *Allington Hunting Party in India.* N.d. Oil on canvas. 91×120 cm. Maidstone Museum and Bentlif Art Gallery, Kent.

Earl, Augustus. "Doña Maria de Jesus." Drawn by Augustus Earl. Engraved by Edward Finden. From Maria Graham, *Journal of a Voyage to Brazil* (London: Longman, 1824).

Erich, August. *Landgrave Maurice of Hesse-Kassel with his Family.* 1618– 1628. Oil on canvas. 230.5 × 422 cm. Museumslandschaft Hessen Kassel, Kassel.

Gainsborough, Thomas. *Mr. and Mrs. Andrews.* Circa 1750. Oil on canvas. 69.8 × 119.4 cm. National Portrait Gallery, London.

Garzoni, Giovanna. *Bitch with Biscuits and Chinese Bowl.* Circa 1648. Oil on canvas. 27.5 × 39.5 cm. Pitti Palace, Florence.

———. *Pug on a Table with Bread, Chinese Tea Bowl, Sopressata and Flies.* Mid-seventeenth century. Watercolor on vellum. 27.3 × 36.8 cm. Doyle New York, New York (January 26, 2015).

Gerhard, Hubert. *Female Pug.* Circa 1600. Cast bronze. 21 cm high. Museumslandschaft Hessen Kassel, Kassel.

Graham, Maria. [M.G.]. "Banian Tree." Etched by James Storer. From Maria Graham, *Journal of a Residence in India*, 2nd ed. (Edinburgh: Constable, 1813).

———. [M.G.]. "Entrance to the Great Cave at Kenary." Etched by James Storer. From Maria Graham, *Journal of a Residence in India*, 2nd ed. (Edinburgh: Constable, 1813).

———. Figures in the Caves at Elephanta. Drawn and etched by M. [Maria] Graham. From Maria Graham, *Journal of a Residence in India*, 2nd ed. (Edinburgh: Constable, 1813).

———. [M.G.]. "Interior of the Great Cave at Carli." Etched by James Storer. From Maria Graham, *Journal of a Residence in India*, 2nd ed. (Edinburgh: Constable, 1813).

———. "Larangeiros." Drawn by Maria Graham. Engraved by Edward Finden. From Maria Graham, *Journal of a Voyage to Brazil* (London: Longman, 1824).

———. "Rio, from the Gloria Hill." Drawn by Maria Graham. Engraved by Edward Finden. From Maria Graham, *Journal of a Voyage to Brazil* (London: Longman, 1824).

———. "San Cristovaõ." Drawn by Maria Graham. Engraved by Edward Finden. From Maria Graham, *Journal of a Voyage to Brazil* (London: Longman, 1824).

———. [M.G.]. "Temple of Maha Deo in Bombay." Etched by James
Storer. From Maria Graham, *Journal of a Residence in India*, 2nd ed.
(Edinburgh: Constable, 1813).

———. [M.G.]. "Temporary Bridge & Bungalo at Barbareen" (tomb of a
Muslim saint). Etched by James Storer. From Maria Graham, *Journal
of a Residence in India*, 2nd ed. (Edinburgh: Constable, 1813).

———. "View of the Corcovado." Drawn by Maria Graham. Engraved by
Edward Finden. From Maria Graham, *Journal of a Voyage to Brazil*
(London: Longman, 1824).

Goltzius, Hendrick. *Justus Lipsius*. 1587. Engraving. 14 × 9.5 cm. National
Gallery of Art, Washington, DC.

———. *Lot and His Daughters*. 1616. Oil on canvas. 140 × 204 cm. Rijksmuseum, Amsterdam.

Hodges, William. *Natives Drawing Water from a Pond with Warren
Hastings' House at Alipur in the Distance*. 1781. Oil on canvas.
69.2 × 91.5 cm. Private collection.

Hogarth, William. *Portrait of a Family*. Circa 1735. Oil on canvas.
53.3 × 74.9 cm. Center for British Art, Paul Mellon Collection.

Hollar, Wenceslaus, after Johannes Nieuhof. "The Supreame Monarch of
the China-Tartarian Empire." In *An Embassy from the East-India
Company of the United Provinces, to the Grand Tartar Cham, Emperor of
China, deliver'd by their excellencies Peter de Goyer and Jacob de Keyzer*,
1673. Etching. Folger Shakespeare Library, Washington, D.C. Call
#141-515f. Plate 1 after page 68.

Höroldt, Johann Gregorius (Meissen porcelain manufactory). *Meissen
Chinoiserie Rinsing Bowl*. 1723–1724. Hard-paste porcelain. 8.5 × 17.7 cm.
CE.74.131. Smithsonian National Museum of American History,
Washington, DC.

———. *Tankard*. Circa 1725. Hard-paste porcelain decorated in underglaze blue. 19.4 × 15.1 × 10.2 cm. 1980.615. Museum of Fine Arts,
Boston.

Kändler, Johann Joachim (Meissen porcelain manufactory). *Pug Bitch with
Puppy [Mops mit Jungtier]* and *Pug Dog [Mops]*. Circa 1741–1745.
Porcelain with enamel paint. 17.9 × 18.6 × 11.8 cm and 17.3 × 19.2 × 11.1 cm.
PE 3894 and 3895. Porzellansammlung, Staatliche Kunstsammlungen
Dresden.

Martin, John. *View of the Temple of Suryah & Fountain of Maha Dao, with a Distant View of North Side of Mansion House.* Engraving with aquatint added by Frederick Christian Lewis after a drawing etched by John Martin. Circa 1819. Plate 6 of *Views of Sezincot* [sic]. [10] leaves of plates: all illustrations (aquatints, hand colored). 44×56 cm. The Lilly Library, Indiana University, Bloomington, IN.

Masanobu, Okumura. *Winter: Pictures of the Four Seasons: Plum Blossoms in the Snow.* 1720s. Woodblock print. Honolulu Museum of Art, Honolulu.

Mauk-Sow-U (?). *E Chaw.* 1772. Colored drawing. Oak Spring Garden Foundation, Upperville, VA (osgf.org).

Miller, John. "Green Tea, Publish'd according to Act of Parliament Dec. 10. 1771." Hand-colored printed frontispiece to John Coakley Lettsom, *The Natural History of the Tea-Tree* (London: Edward and Charles Dilly, 1772). Wellcome Library, London.

Oudry, Jean-Baptiste. *The Seraglio of the Pug.* 1734. Oil on canvas. 105.5×135 cm. Audap et Mirabaud, Paris.

Potter, Paulus. *A Hall Interior with a Group of Eight Dogs, Two Seated upon a Green Cushion, the Others Resting and Playing on the Ground.* 1649. Oil on canvas. 119×150 cm. Noortman Master Paintings, Amsterdam.

Pouncy, Benjamin Thomas. *Banyan Tree, Engraving after William Hodges.* Plate 5 of William Hodges's *Travels in India, during the years 1780, 1781, 1782, & 1783* published in London in 1793. Reprinted in *British Encounters with India, 1750–1830: A Sourcebook,* edited by Tim Keirn and Norbert Schürer, 211. Basingstoke: Palgrave Macmillan, 2011.

Rottenhammer, Hans. *Minerva with the Muses on Mount Helicon.* 1603. Oil on canvas. 186×308.8 cm. Germanisches Nationalmuseum, Nuremberg.

Sandby, Paul. *The Analyst Besh[itte]n in his Own Taste.* 1753. Etching. 26.4×183 cm. British Museum, London.

Vandergucht, G[erard]., after C[lement]. Lemprière. Thomas Wright, *Machina Coelestis, or the Great Orrery.* 1730. Engraving. 20×28 inches. Science Museum Group Collection, London.

Wright of Derby, Joseph. *A Philosopher Giving That Lecture on the Orrery, in Which a Lamp Is Put in Place of the Sun.* Circa 1766. Oil on canvas. 147.2×203.2 cm. Derby Museum and Art Gallery.

Zoffany, Johan. *Warren Hastings and His Second Wife in Their Garden at Alipore*. Circa 1784–1787. Oil on canvas. 90.2 × 119.4 cm. Collection of the Victoria Memorial Hall, Kolkata, India.

Zhou Fang. *Ladies Wearing Flowers in Their Hair*. Late eighth century. Ink and color on silk handscroll. 46 × 180 cm. Liaoning Provincial Museum, Shenyang.

Manuscripts

Blake, John Bradby. "Notes towards a Natural History of China" and personal correspondence. M–152 (multiple volumes). Oak Spring Garden Foundation, Upperville, VA.

British Government. Papers relating to tea. Chatham Papers (1786–1800), PRO 30/8/294; Treasury Papers (1778), T 1/542. National Archives, Kew.

Cuninghame, James. Notes on Botany and Zoology in East Asia and the Canary Islands. 1698–1699. Sloane 2376. British Library, London.

Cuninghame, James, Hans Sloane, and James Petiver. Correspondence. Sloane MSS 3321, 3334, 4025. British Library, London.

East India Company. Minute books and outgoing correspondence. India Office Records, B/4, B/43, E/3/94, E/4/863, IOR/G/12/6, 910. British Library, London.

Ellis, John. Correspondence and memoranda. John Ellis Note-Book 2. MS/292. Linnean Society of London.

Macartney, Sir George. "Mission to China, 1792–1794." Charles W. Wason Asia Collections, Kroch Rare Book and Manuscript Library, Cornell University, Ithaca, NY. 10 vols. MSS DS117.

———. "Commonplace Book kept by Lord Macartney during his Embassy in China, 1792." Wason MSS DS. M118.

———. Macartney Papers. Twelve envelopes. 貴重書 MS–42. Toyo Bunko Oriental Library, Tokyo.

Printed Works

Adams, Nöel. "Between Myth and Reality: Hunter and Prey in Early Anglo-Saxon Art." In *Representing Beasts in Early Medieval England*

and Scandinavia, edited by Michael D. J. Bintley and Thomas J. T. Williams, 13–52. Woodbridge: Boydell and Brewer, 2015.

Abercrombie, John. *The Propagation and Botanical Arrangements of Plants and Trees, Useful and Ornamental, Proper for Cultivation in Every Department of Gardening.* 2 vols. London, 1784.

Addison, Joseph, and Richard Steele. *The Guardian.* Edited by John Calhoun Stephens. Lexington: University of Kentucky Press, 1982.

———. *The Spectator.* Edited by Donald F. Bond. 5 vols. Oxford: Clarendon, 1965.

———. *The Tatler.* Edited by Donald F. Bond. 3 vols. Oxford: Clarendon, 1987.

"The Age of Networks: Social, Cultural, and Technological Connections" (2006). University of Illinois, Urbana-Champaign. http://www.ncsa .illinois.edu/Conferences/Networks/.

Aiton, William. *Hortus Kewensis; Or, A Catalogue of the Plants Cultivated in the Royal Botanic Garden at Kew. By William Aiton, Gardener to His Majesty.* 3 vols. London, 1789.

Allan, John. "Some Post-Medieval Documentary Evidence for the Trade in Ceramics." In *Ceramics and Trade: The Production and Distribution of Later Medieval Pottery in North-West Europe*, edited by Peter Davey and Richard Hodges, 37–45. Sheffield: University of Sheffield Department of Prehistory and Archaeology, 1983.

Anderson, Benedict. *Imagined Communities: Reflections on the Origin and Spread of Nationalism.* London: Verso, 1991.

Andrews, A. D. "Cyclopaedia of Telescope Makers." *Irish Astronomical Journal* 20, no. 3 (March 1992): 176.

Archer, Mildred. "The Daniells in India and Their Influence on British Architecture." *RIBA Journal* 67 (1960): 439–444.

———. *Early Views of India: The Picturesque Journey of Thomas and William Daniell, 1786–1794.* London: Thames and Hudson, 1980.

Aubin, Penelope. *A Collection of Entertaining Histories and Novels, Designed to Promote the Cause of Virtue and Honour.* 3 vols. London, 1739.

Baines, Paul. "Alexander Pope." In *The Cambridge Companion to English Poets*, edited by Claude Rawson, 235–253. Cambridge: Cambridge University Press, 2011.

Baird, Ileana, ed. *Social Networks in the Long Eighteenth Century: Clubs, Literary Salons, Textual Coteries.* Newcastle upon Tyne: Cambridge Scholars, 2014.

Baird, Ileana, and Christina Ionescu, eds. *Eighteenth-Century Thing Theory in a Global Context: From Consumerism to Celebrity Culture.* Farnham: Ashgate, 2013.

Ballantyne, Tony. *Orientalism and Race: Aryanism in the British Empire.* Basingstoke: Palgrave, 2002.

Banks, Sir Joseph. *The Indian and Pacific Correspondence of Sir Joseph Banks, 1768–1720.* Edited by Neil A. Chambers. 8 vols. London: Pickering and Chatto, 2008–2014.

Bannet, Eve Tavor. *Transatlantic Stories and the History of Reading, 1720–1810: Migrant Fictions.* Cambridge: Cambridge University Press, 2011.

Barabási, Albert-László. *Bursts: The Hidden Patterns behind Everything We Do.* New York: Dutton, 2010.

———. *Linked: The New Science of Networks.* Cambridge, MA: Perseus, 2002.

Barabási, Albert-László, and Réka Albert. "Emergence of Scaling in Random Physics Networks." *Science* 286 (2013): 509–512.

Barbauld, Anna Letitia. *Selected Poetry and Prose.* Edited by William McCarthy and Elizabeth Kraft. Peterborough, ON: Broadview, 2002.

Barnett, Suzanne W. "Silent Evangelism: Presbyterians and the Mission Press, 1807–1860." *Journal of Presbyterian History (1962–1985)* 49, no. 4 (1971): 287–302.

Barrow, John. *Travels in China: Containing Descriptions, Observations and Comparisons, Made and Collected in the Course of a Short Residence at the Imperial Palace of Yuen-Min-Yuen, and on the Subsequent Journey through the Country from Pekin to Canton* (1804). Cambridge: Cambridge University Press, 2010.

Beach, Adam R. "Aubin's *The Noble Slaves,* Montagu's Spanish Lady, and English Feminist Writing about Sexual Slavery in the Ottoman World." *Eighteenth-Century Fiction* 29, no. 4 (Summer 2017): 583–606.

Beattie, James. "Eco-Cultural Networks in Southern China and Colonial New Zealand: Cantonese Market Gardening and Environmental Exchange 1860s–1910s. In *Eco-Cultural Networks and the British*

Empire: New Views on Environmental History, edited by James Beattie, Edward Melillo, and Emily O'Gorman, 151–179. London: Bloomsbury, 2015.

Beattie, James, Edward Melillo, and Emily O'Gorman, eds. *Eco-Cultural Networks and the British Empire: New Views on Environmental History*. London: Bloomsbury, 2015.

Beattie, James, Edward Melillo, and Emily O'Gorman. "Introduction." In *Eco-Cultural Networks and the British Empire: New View on Environmental History*, edited by James Beattie, Edward Melillo, and Emily O'Gorman, 3–20. London: Bloomsbury, 2015.

———. "Rethinking the British Empire through Eco-Cultural Networks: Materialist-Cultural Environmental History, Relational Connections and Agency." *Environment and History* 20, no. 4 (2014): 561–575.

Beevers, David. "'Mand'rin only is the man of taste': 17th and 18th Century Chinoiserie in Britain." In *Chinese Whispers: Chinoiserie in Britain, 1650–1930*, edited by David Beevers, 13–26. Brighton: Royal Pavilion and Museums, 2008.

Bell, John. *Travels from St. Petersburgh in Russia to Various Parts of Asia*. 2 vols. Edinburgh: William Creech, 1788.

Berg, Maxine. "Consumption in Eighteenth- and Early Nineteenth-Century Britain." In *Industrialisation, 1700–1860*, edited by Roderick Floud and Paul Johnson, 357–386. Vol. 1 of *The Cambridge Economic History of Modern Britain*. Cambridge: Cambridge University Press, 2004.

———. *Luxury and Pleasure in Eighteenth-Century Britain*. Oxford: Oxford University Press, 2005.

Bergstrom, Carson. "Literary Coteries, Network Theory, and the Literary and Philosophical Society of Manchester." *ANQ: A Quarterly Journal of Short Articles, Notes, and Reviews* 26, no. 3 (2013): 180–188.

Blakesley, David, and Thomas Rickert. "From Nodes to Nets: Our Emerging Culture of Complex Interactive Networks." *JAC: A Journal of Composition Theory* 24, no. 4 (2004): 821–830.

Bolton, Kingsley. Introduction to *A Vocabulary of the Canton Dialect* by Robert Morrison, v–xliv. London: Ganesha; Tokyo: Edition Synapse, 2001.

Bourne, Val. "Sezincote Garden, Gloucestershire: A Passage to India." *The Telegraph*, July 6, 2012. https://www.telegraph.co.uk/gardening/9379639/Sezincote-garden-Gloucestershire-a-passage-to-India.html.

Bowen, H. V. *The Business of Empire: The East India Company and Imperial Britain, 1756–1833.* Cambridge: Cambridge University Press, 2006.

Brewer, John, and Roy Porter, eds. *Consumption and the World of Goods.* London: Routledge, 1993.

Brookes, Richard. *The Natural History of Quadrupedes, Including Amphibious Animals, Frogs, and Lizards: With Their Properties and Uses in Medicine.* 6 vols. London: J. Newbery, 1763.

Brown, Hilary, and Gillian Dow, eds. *Readers, Writers, Salonnières: Female Networks in Europe, 1700–1900.* New York: Peter Lang, 2011.

Caballero, M. Soledad. "'For the Honour of Our Country': Maria Dundas Graham and the Romance of Benign Domination." *Studies in Travel Writing* 9 (2005): 111–131.

Cams, Mario. "Not Just a Jesuit Atlas of China: Qing Imperial Cartography and Its European Connections." *Imago Mundi* 69, no. 2 (2017): 188–201.

Carey, Eustace. *Memoir of William Carey, D.D.* London: Jackson and Walford, 1836.

Carlton, Charles, and Caroline Carlton. "Gardens of the Raj." *History Today* 46, no. 7 (1996): 22–28.

Casson, Mark. "Networks in Economic and Business History: A Theoretical Perspective." In *Cosmopolitan Networks in Commerce and Society, 1660–1914,* edited by Andreas Gestrich and Margrit Schulte Beerbühl, 17–49. London: German Historical Institute, 2011.

Castells, Manuel. *The Information Age: Economy, Society and Culture.* Vol. 1, *The Rise of the Network Society.* Cambridge, MA: Blackwell, 1996.

Chatterjee, KumKum, and Clement Hawes, eds. *Europe Observed: Multiple Gazes in Early Modern Encounters.* Lewisburg, PA: Bucknell University Press, 2008.

Classen, Albrecht. "A Global Epistolary Network: Eighteenth-Century Jesuit Missionaries Writing Home with an Emphasis on Philipp Segesser's Correspondence from Sonora/Mexico." *Studia Neophilologica* 86 (2014): 79–94.

Clingham, Greg. "Cultural Difference in George Macartney's *An Embassy to China, 1792–94.*" *Eighteenth-Century Life* 39, no. 2 (2015): 1–29.

———. "Johnson and China: Culture, Commerce, and the Dream of the Orient in Mid-Eighteenth-Century England." *1650–1850: Ideas, Aesthetics, and Inquiries in the Early Modern Era* 24 (2019): 178–242.

Codell, Julie F. "The Art of Transculturation." In *Transculturation in British Art, 1770–1930*, edited by Julie F. Codell, 1–20. Farnham: Ashgate, 2012.

Codr, Dwight. *Raving at Usurers: Anti-Finance and the Ethics of Uncertainty in England, 1690–1750*. Charlottesville and London: University of Virginia Press, 2016.

Cohen, David. "All the World's a Net." *New Scientist* 174, no. 2338 (April 13, 2002). https://www.newscientist.com/article/mg17423384-700 -all-the-worlds-a-net/.

Coleridge, Samuel Taylor. *The Complete Poems*. Edited by William Keach. Harmondsworth: Penguin, 1997.

Collins, Harry M., and Steven Yearley. "Epistemological Chicken." In *Science as Practice and Culture*, edited by Andrew Pickering, 301–326. Chicago: University of Chicago Press, 1992.

Conner, Patrick. *Oriental Architecture in the West*. London: Thames and Hudson, 1979.

Cooper, Rose, and Darcy White. "Teaching Transculturation: Pedagogical Processes." *Journal of Design History* 18, no. 3 (2005): 285–292.

Cope, Kevin L., and Samara Anne Cahill, eds. *Citizens of the World: Adapting in the Eighteenth Century*. Lewisburg, PA: Bucknell University Press, 2015.

Coventry, Francis. *The History of Pompey the Little: Or, the Life and Adventures of a Lap-Dog*. London: M. Cooper, 1751.

Cowan, Brian. *The Social Life of Coffee: The Emergence of the British Coffeehouse*. New Haven, CT: Yale University Press, 2005.

Cranmer-Byng, J. L. "Lord Macartney's Embassy to Peking in 1793, from Official Chinese Documents." *Journal of Oriental Studies* 4, no. 1–2 (1957–58): 117–187.

Crosbie, Barry. *Irish Imperial Networks: Migration, Social Communication and Exchange in Nineteenth-Century India*. Cambridge: Cambridge University Press, 2012.

Cuninghame, James. "Part of Two Letters to the Publisher from Mr James Cunningham, F.R.S. and Physician to the English at Chusan in China." *Philosophical Transactions of the Royal Society of London* 23 (1702–1703): 1201–1209.

Cushion, John P. *Animals in Pottery and Porcelain*. London: Studio Vista, 1974.

Czennia, Bärbel. "Nationale und kulturelle Identitätsbildung in Grossbritannien, 1660–1750. Eine historische Verlaufsbeschreibung," In *Muster und Funktionen kultureller Selbst- und Fremdwahrnehmung. Beiträge zur internationalen Geschichte der sprachlichen und literarischen Emanzipation*, edited by Ulrike-Christine Sander and Fritz Paul, 355–390. Göttingen: Wallstein, 2000.

Daily Advertiser, November 2, 1772.

Daily, Christopher A. *Robert Morrison and the Protestant Plan for China*. Hong Kong: Hong Kong University Press, 2013.

Dalrymple, Alexander, ed. *Oriental Repertory*. 2 vols. London: East India Company, 1791.

Damodaran, Vinita, Anna Winterbottom, and Alan Lester, eds. *The East India Company and the Natural World*. Houndmills: Palgrave Macmillan, 2015.

Daniell, Thomas, and William Daniell. *Oriental Scenery: 150 Views of Architecture, Antiquities and Landscape Scenery of Hindostan*. London, 1795–1816.

Daniels, Stephen. *Humphry Repton: Landscape Gardening and the Geography of Georgian England*. New Haven, CT: Yale University Press, 1999.

Dapper, Olfert. "A Third Embassy to the Emperor of China and East-Tartary, Under the Conduct of the Lord Pieter van Hoorn, Containing Several Remarks in Their Journey through the Provinces of Fokien, Chekiang, Xantung, and Nanking, to the Imperial Court at Peking." In *Atlas Chinensis: Being a second part of a relation of remarkable passages in two embassies from the East-India Company of the United Provinces, to the Vice-Roy Singlamong and General Taising Lipovi, and to Konchi, Emperor of China and East-Tartary: with a relation of the Netherlanders assisting the Tartar against Coxinga, and the Chinese fleet, who till then were masters of the sea, and a more exact geographical description than formerly, both of the whole empire of China in general, and in particular of every of the fifteen provinces*, translated by John Ogilby, vol. 2, 203–723. London: John Ogilby, 1671.

Darnton, Eric. *Poetry and the Police: Communication Networks in Eighteenth-Century Paris*. Cambridge, MA: Belknap Press of Harvard University Press, 2010.

Dart, Gregory. *Metropolitan Art and Literature, 1810–1840: Cockney Adventures*. Cambridge: Cambridge University Press, 2012.

Davies, Simon, Daniel Sanjiv Roberts, and Gabriel Sánchez Espinosa. *India and Europe in the Global Eighteenth Century*. Oxford: Voltaire Foundation, 2014.

Davies, Timothy. "Trading Letters in the Arabian Seas: The Correspondence Networks of British Merchants in Eighteenth-Century Western India." In special issue, "Transcultural Networks in the Indian Ocean, Sixteenth to Eighteenth Centuries: Europeans and Indian Ocean Societies in Interaction," edited by Su Fang Ng. *Forms of Discourse and Culture* 48, no. 2 (2015): 215–236.

Daw, Margery [pseud.]. "Literary Lady." *British Stage and Literary Cabinet* 5, no. 59 (1821): 381.

De Almeida, Hermione, and George H. Gilpin. *Indian Renaissance: British Romantic Art and the Prospect of India*. London: Ashgate, 2005.

[Defoe, Daniel]. *The Fortunes and Misfortunes of the Famous Moll Flanders &c.* London: W. Chetwood and T. Edling, 1721.

Delbourgo, James. *Collecting the World: Hans Sloane and the Origins of the British Museum*. London: Allen Lane, 2017.

Desmond, Ray. *The European Discovery of the Indian Flora*. Oxford: Oxford University Press, 1992.

De Zwart, Pim. "Globalization in the Early Modern Era: New Evidence from the Dutch-Asiatic Trade, c. 1600–1800." *Journal of Economic History* 76, no. 2 (2016): 520–558.

Dickens, Charles. *Great Expectations*. Edited by David Trotter. Harmondsworth: Penguin, 1996.

Dirks, Nicholas B. *The Scandal of Empire: India and the Creation of Imperial Britain*. Cambridge, MA: Belknap Press of Harvard University Press, 2006.

Dodsley, Robert. *The Toy-Shop. To which are added, Poems and Epistles on Several Occasions*. 11th ed. London: Robert Dodsley, 1744.

Donne, John. "Satire III." *Representative Poetry Online*. University of Toronto Libraries. https://rpo.library.utoronto.ca/poems/satire-iii.

Douglas, R. K. "Morrison, Robert (1782–1834)." Revised by Robert Bickers. *Oxford Dictionary of National Biography*. Oxford: Oxford University Press,

2004. Online edition, May 2007. https://www.oxforddnb.com/view/10
.1093/ref:odnb/9780198614128.001.0001/odnb-9780198614128-e-19330.

Drayton, Richard. *Nature's Government: Science, Imperial Britain, and the "Improvement" of the World*. New Haven, CT: Yale University Press, 2000.

Dyer, John. *The Fleece. A Poem. In Four Books*. London: R. and J. Dodsley, 1757.

Ellis, John. *Directions for Bringing over Seeds and Plants from the East-Indies and Other Distant Countries, in a State of Vegetation*. London: L. Davis, 1770.

———. *Some Additional Observations on the Method of Preserving Seeds from Foreign Parts, for the Benefit of Our American Colonies*. London: W. Bowyer and J. Nichols, 1773.

Ellis, Markman. *The Coffee House: A Cultural History*. London: Weidenfeld and Nicolson, 2004.

———, ed. *Eighteenth-Century Coffee-House Culture*. 4 vols. London: Pickering and Chatto, 2006.

Ellis, Markman, Richard Coulton, and Matthew Mauger. *Empire of Tea: The Asian Leaf that Conquered the World*. London: Reaktion, 2015.

Ellis, Markman, Richard Coulton, Matthew Mauger, et al., eds. *Tea and the Tea-Table in Eighteenth-Century England*. 4 vols. London: Pickering and Chatto, 2010.

Fang, Karen. "Empire, Coleridge, and Charles Lamb's Consumer Imagination." *Studies in English Literature, 1500–1900* 43 (2003): 815–843.

Farrell, William. "Smuggling Silks into Eighteenth-Century Britain: Geography, Perpetrators, and Consumers." *Journal of British Studies* 55, no. 2 (2016): 268–294.

Feinstein, Charles. "Pessimism Perpetuated: Real Wages and the Standard of Living in Britain during and after the Industrial Revolution." *Journal of Economic History* 58 (1998): 625–658.

Ferguson, Patricia F. "Sprimont's Complaint: Buying and Selling Continental Porcelain in London (1730–1753)." In *Arts Antiques London Catalogue: 2012*, 11–21. London: Haughton International Fairs, 2012.

Ferrier, Susan. *Marriage*. Edited by Herbert Foltinek. Oxford: Oxford University Press, 2001.

Fisher, Michael H., ed. *The Travels of Dean Mahomet: An Eighteenth-Century Journey through India* (1794). Berkeley and Los Angeles: University of California Press, 1997.

——, ed. *Visions of Mughal India: An Anthology of European Travel Writing*. London: I. B. Tauris, 2007.

Floud, Roderick, and Paul Johnson, eds. *The Cambridge Economic History of Modern Britain*. 3 vols. Cambridge: Cambridge University Press, 2004.

Forbes, James. *Oriental Memoirs: Selected and Abridged from a Series of Familiar Letters Written During Seventeen Years Residence in India: Including Observations on Parts of Africa and South America, and a Narrative of Occurrences in Four India Voyages; Illustrated by Engravings from Original Drawings*. 4 vols. London: White, Cochrane, and Co., 1813.

Fort St. George. "The Eighth Book of East India Plants, Sent from Fort St George to Mr James Petiver Apothecary, and F. R. S. with His Remarks on Them." *Philosophical Transactions of the Royal Society of London* 23 (1702–1703): 1450–1460.

Foucault, Michel. "The Order of Discourse." In *Untying the Text: A Post-Structuralist Reader*, edited by Robert Young, 51–78. Trans. Ian McLeod. London: Routledge, 1981.

Freeman, Linton C. *The Development of Social Network Analysis: A Study in the Sociology of Science*. Vancouver, BC: Empirical, 2004.

Galloway, Alexander R., and Eugene Thacker. *The Exploit: A Theory of Networks*. Minneapolis: University of Minnesota Press, 2007.

Garcia, Humberto. *Islam and the English Enlightenment, 1670–1840*. Baltimore, MD: Johns Hopkins University Press, 2011.

Garway, Thomas. *An Exact Description of the Growth, Quality, and Vertues of the Leaf Tee, alias Tay*. [London], [c. 1664].

Gestrich, Andreas, and Margrit Schulte Beerbühl, eds. *Cosmopolitan Networks in Commerce and Society, 1660–1914*. London: German Historical Institute, 2011.

Gilbert, Felix. "The New Diplomacy of the Eighteenth Century." *World Politics* 4, no. 1 (October 1951): 1–38.

Goffman, Daniel. *The Ottoman Empire and Early Modern Europe*. Cambridge: Cambridge University Press, 2002.

Goldsmith, Oliver. *Collected Works of Oliver Goldsmith*. Edited by Arthur Friedman. 5 vols. Oxford: Clarendon, 1966.

Golinski, Jan. "Astronomy." In *The Blackwell Companion to the Enlightenment*, edited by John Yolton, Roy Porter, Pat Rogers, and Barbara Stafford, 44–45. Cambridge, MA: Blackwell, 1991.

Gollapudi, Aparna. "Virtuous Voyages in Penelope Aubin's Fiction." *Studies in English Literature, 1500–1900* 45, no. 3 (Summer 2005): 669–690.

Gottlieb, Evan, ed. *Global Romanticism: Origins, Orientations, and Engagements, 1760–1820*. Lewisburg, PA: Bucknell University Press, 2015.

Gottlieb, Evan. *Romantic Globalism: British Literature and Modern World Order, 1750–1830*. Columbus: Ohio State University Press, 2014.

Graham, Maria. *Journal of a Residence in Chile During the Year 1822*. New York: Praeger, 1969.

———. *Journal of a Residence in India*. 2nd ed. Edinburgh: Constable, 1813.

———. *Journal of a Voyage to Brazil, and Residence There, during Part of the Years 1821, 1822, 1823*. London: Longman, 1824.

———. *Journal of a Voyage to Brazil, and Residence There, during Part of the Years 1821, 1822, 1823*. New York: Praeger, 1969.

———. *Three Months Passed in the Mountains East of Rome*. London: Longman, 1820.

Graves, Richard. "The Insolence of Office." In *Euphrosyne: Or, the Amusements of the Road of Life*, 88–89. London: James Dodsley, 1776.

Grove, Richard H. *Green Imperialism: Colonial Expansion, Tropical Island Edens and the Origins of Environmentalism, 1600–1860*. Cambridge: Cambridge University Press, 1995.

Guerra, Lucia. "Nociones de la otredad en los textos de viajeras del siglo XIX." *Chasqui: Revista de Literatura Latinoamericana* 45, no. 1 (May 2016): 30–41.

Haglund, Betty. "The Botanical Writings of Maria Graham." *Journal of Literature and Science* 4, no. 1 (2011): 44–58.

Hall, Catherine, and Sonya O. Rose, eds. *At Home with the Empire: Metropolitan Culture and the Imperial World*. Cambridge: Cambridge University Press, 2006.

Haraway, Donna. *The Companion Species Manifesto: Dogs, People, and Significant Otherness*. Chicago: Prickly Paradigm, 2003.

Harrison-Hall, Jessica. "A Meeting of East and West: Print Sources for Eighteenth-Century Chinese Trade Porcelain." *Apollo: The International Magazine of the Arts* 384 (February 1994): 3–6.

Halpern, Linda Cabe. "The Uses of Paintings in Garden History." In *Garden History, Issues, Approaches, Methods*, edited by John Dixon Hunt, 183–202. Washington, DC: Dumbarton Oaks Research Library and Collection, 1992.

Hargrave, Jennifer L. "*Marco Polo* and the Emergence of British Sinology." *SEL: Studies in English Literature, 1500–1900* 56, no. 3 (2016): 515–537.

Hawkes, David, ed. and trans. *The Songs of the South: An Ancient Chinese Anthology of Poems by Qu Yuan and Other Poets*. Harmondsworth: Penguin, 1985.

Hayward, Jennifer. "'The Uncertainty of the End Keeps up the Interest': Maria Graham's *Journal of a Residence in Chile* as Life Writing." *Auto/Biography Studies* 17, no. 1 (2001): 43–64.

Haywood, Eliza. *The Invisible Spy*. 4 vols. London: T. Gardner, 1755.

Hazlitt, William. *Metropolitan Writings*. Edited by Gregory Dart. Manchester: Carcanet, 2005.

Head, Raymond. *The Indian Style*. London: Allen and Unwin, 1986.

Hegel, Georg W. F. *Lectures on the Philosophy of World History. Introduction: Reason in History*. Edited by Johannes Hoffmeister. Translated by H. B. Nisbet. Cambridge: Cambridge University Press, 1975.

Herbert, Eugenia W. *Flora's Empire: British Gardens in India*. Philadelphia: University of Pennsylvania Press, 2011.

———. "Peradeniya and the Plantation Raj in Nineteenth-Century Ceylon." In *Eco-Cultural Networks and the British Empire: New Views on Environmental History*, edited by James Beattie, Edward Melillo, and Emily O'Gorman, 123–150. London: Bloomsbury, 2015.

Herschel, William. "On the Construction of the Heavens" (1785). *Philosophical Transactions of the Royal Society of London* 75 (1785): 213–266.

Hess-Lüttich, Ernest W. B., and Yoshito Takahashi, eds. *Orient im Okzident—Okzident im Orient: West-Östliche Begegnungen in Sprache und Kultur, Literatur und Wissenschaft*. Frankfurt: Peter Lang, 2015.

Hevia, James L. *Cherishing Men from Afar: Qing Guest Ritual and the Macartney Embassy of 1793*. Durham, NC: Duke University Press, 1999.

———. *English Lessons: The Pedagogy of Imperialism in Nineteenth-Century China*. Durham, NC: Duke University Press; Hong Kong: Hong Kong University Press, 2003.

Higgins, David. *Romantic Englishness: Local, National, and Global Selves, 1780–1850*. Basingstoke: Palgrave Macmillan, 2014.

Historic Buildings and Monuments Commission. "Daylesford House" [list entry number: 1000760]. https://www.historicengland.org.uk/listing/the-list/list-entry/1000760.

Hobhouse, Penelope. *Gardening through the Ages: An Illustrated History of Plants and Their Influence on Garden Styles from Ancient Egypt to the Present Day*. New York: Barnes and Noble, 1997.

Hoock, Holger. *Empires of the Imagination: Politics, War, and the Arts in the British World*. London: Profile, 2010.

Hoskin, Michael. *The Construction of the Heavens: William Herschel's Cosmology*. Cambridge: Cambridge University Press, 2012.

———. *William and Caroline Herschel: Pioneers in Late 18th-Century Astronomy*. London: Springer, 2014.

Howard, David, and John Ayers. *China for the West: Chinese Porcelain and Other Decorative Arts for Export Illustrated from the Mottahedeh Collection*. 2 vols. London and New York: Sotheby Parke Bernet, 1978.

Howe, Eunice. "Architecture in Vasari's 'Massacre of the Huguenots.'" *Journal of the Warburg and Courtauld Institutes* 39 (1976): 258–261.

Hsia, R. Po-Chia. *A Jesuit in the Forbidden City: Matteo Ricci, 1552–1610*. Oxford: Oxford University Press, 2010.

Hume, David. *Essays: Moral, Political and Literary*. Oxford: Oxford University Press, 1963.

Hull, Simon P. *Charles Lamb, Elia and the London Magazine*. London: Pickering and Chatto, 2010.

Hunt, Leigh. "India." *The Examiner* 560 (September 20, 1818): 593–594.

Hussey, Christopher. *English Country Houses: Late Georgian 1800–1840*. London: Country Life, 1958.

Irwin, Robert. *Dangerous Knowledge: Orientalism and Its Discontents*. Woodstock and New York: Overlook, 2006.

Jarvis, Charles E., and Philip H. Oswald. "The Collecting Activities of James Cuninghame FRS on the Voyage of Tuscan to China (Amoy) between 1697 and 1699." *Notes and Records of the Royal Society* 69 (2015): 135–153.

Jensen, Niklas Thode. "The Tranquebarian Society: Science, Enlightenment and Useful Knowledge in the Danish-Norwegian East Indies, 1768–1813." *Scandinavian Journal of History* 40, no. 4 (2015): 535–561.

Johanyak, Debra, and Walter S. H. Lim, eds. *The English Renaissance, Orientalism, and the Idea of Asia.* New York: Palgrave Macmillan, 2010.

Johnson, Samuel. *A Dictionary of the English Language.* 2 vols. London: J. and P. Knapton, 1755.

———. *The Works of Samuel Johnson.* Vol. 20, *Johnson on Demand: Reviews, Prefaces, and Ghost-Writing,* edited by O M Brack, Jr. and Robert DeMaria, Jr., "Preface to Richard Rolt, *A New Dictionary of Trade and Commerce* (1756)."

Jones, Sir William. *A Discourse on the Institution of a Society for Enquiring into the History, Civil and Natural, the Antiquities, Arts, Sciences, and Literature of Asia.* London: T. Payne, 1784.

———. *Letters of Sir William Jones.* Edited by Garland Cannon. 2 vols. Oxford: Clarendon, 1970.

Kammerschmidt, Valentin, and Joachim Wilke. *Die Entdeckung der Landschaft: Englische Gärten des 18. Jahrhunderts.* Stuttgart: Deutsche Verlagsanstalt, 1990.

Kändler, Johann Joachim. *Die Arbeitsberichte des Meissener Porzellanmodelleurs Johann Joachim Kaendler, 1706–1775.* Edited by Ulrich Pietsch. Leipzig: Edition Leipzig, 2002.

Keirn, Tim, and Norbert Schürer, eds. *British Encounters with India, 1750–1830: A Sourcebook.* Basingstoke: Palgrave Macmillan, 2011.

Kejariwal, O. P. *The Asiatic Society of Bengal and the Discovery of India's Past.* Delhi: Oxford University Press, 1988.

Kilpatrick, Jane. *Gifts from the Gardens of China: The Introduction of Traditional Chinese Garden Plants to Britain, 1698–1862.* London: Frances Lincoln, 2007.

King, Henry C. (in collaboration with John R. Millburn). *Geared to the Stars: The Evolution of Planetariums, Orreries and Astronomical Clocks.* Toronto: University of Toronto Press, 1978.

Kircher, Athanasius. *China monumentis, qua sacris qua profanis, nec non variis naturae & artis spectaculis, aliarumque rerum memorabilium argumentis illustrata.* Amsterdam: Jacob van Meurs, 1667.

———. "Some Special Remarks Taken out of Athanasius Kircher's Antiquities of China." In *An Embassy from the East-India Company of*

the United Provinces, to the Grand Tartar Cham, Emperor of China, by
Johan Nieuhof, edited and translated by John Ogilby, 319–431. London:
John Ogilby, 1673.

Kitson, Peter J. "'Bid him bow down to that which is above him': The
'Kowtow Controversy' and the Representations of Asian Ceremonials
in Romantic Literature." In *Coleridge, Romanticism and the Orient:
Cultural Negotiations*, edited by David Vallins, Kaz Oishi, and Seamus
Perry, 19–37. London: Bloomsbury, 2013.

———. *Forging Romantic China: Sino-British Cultural Exchange, 1760–
1840*. Cambridge: Cambridge University Press, 2013.

Kitson, Peter J., and Robert Markley, eds. *Writing China: Essays on the
Amherst Embassy (1816) and Sino-British Cultural Relations*. Cambridge:
D. S. Brewer, 2016.

Klancher, Jon. "Discriminations, or Romantic Cosmopolitanisms in
London." In *Romantic Metropolis: The Urban Scene of British Culture,
1780–1840*, edited by James Chandler and Kevin Gilmartin, 65–84.
Cambridge: Cambridge University Press, 2005.

Klekar, Cynthia. "'Prisoners in Silken Bonds': Obligation, Trade, and
Diplomacy in English Voyages to Japan and China." *Journal for Early
Modern Cultural Studies* 6, no. 1 (2006): 84–105.

Koch, Ute Christina. "Count Brühl and His Collection of Porcelain
Boxes." In *Going for Gold: Craftsmanship and Collecting of Gold Boxes*,
edited by Tessa Murdoch and Heike Zech, 184–194. Brighton: Sussex
Academic Press, 2014.

Koerner, Lisbet. *Linnaeus: Nature and Nation*. Cambridge, MA: Harvard
University Press, 1999.

Kowaleski-Wallace, Elizabeth. *Consuming Subjects: Women, Shopping, and
Business in the Eighteenth Century*. New York: Columbia University
Press, 1997.

———. "Women, China, and Consumer Culture in Eighteenth-Century
England." *Eighteenth-Century Studies* 29, no. 2 (1995): 153–167.

Kozaczka, Edward. "Penelope Aubin and Narratives of Empire."
Eighteenth-Century Fiction 25, no. 1 (Fall 2012): 199–225.

Krelage, J. H. "On Polyanthus Narcissi." *Journal of the Royal Horticultural
Society* 12 (1890): 339–346.

Lamb, Charles. *Elia and The Last Essays of Elia*. Edited by Jonathan Bate. Oxford: Oxford University Press, 1987.

Lamb, Charles, and Mary Lamb. *The Works of Charles and Mary Lamb*. Edited by E. V. Lucas. 6 vols. London: Methuen, 1905.

Landes, David S. *Revolution in Time: Clocks and the Making of the Modern World*. Cambridge, MA: Harvard University Press, 1983.

Langdon, William B. *Ten Thousand Chinese Things. A Descriptive Catalogue of the Chinese Collection, Now Exhibiting at St. George's Place, Hyde Park Corner, London, with Condensed Accounts of the Genius, Government, History, Literature, Agriculture, Arts, Trade, Manners, Customs, and Social Life of the People of the Celestial Empire*. 12th English ed. London, 1842.

Larson, Greger, Elinor K. Karlsson, Angela Perri, et al. "Rethinking Dog Domestication by Integrating Genetics, Archeology, and Biogeography." *Proceedings of the National Academy of Sciences of the United States of America* 109, no. 23 (2012): 8878–8883.

Latour, Bruno. *Aramis, ou, L'amour des techniques*. Paris: La Découverte, 1992. Translated by Catherine Porter as *Aramis, or, The Love of Technology*. Cambridge, MA: Harvard University Press, 1996.

———. *Les microbes: Guerre et paix, suivi de irréductions*. Paris: Editions A. M. Métailié, 1984. Translated by Alan Sheridan and John Law as *The Pasteurization of France*. Cambridge, MA: Harvard University Press, 1988.

———. *Reassembling the Social: An Introduction to Actor-Network-Theory*. Oxford: Oxford University Press, 2005.

Lawson, Joseph. "The Chinese State and Agriculture in an Age of Global Empires, 1880–1949." In *Eco-Cultural Networks and the British Empire: New Views on Environmental History*, edited by James Beattie, Edward Melillo, and Emily O'Gorman, 44–67. London: Bloomsbury, 2015.

Le Corbeiller, Clare. "German Porcelain of the Eighteenth Century." *Metropolitan Museum of Art Bulletin* 47, no. 4 (1990): 1–55.

Leclerc, Georges-Louis. *Histoire Naturelle par Buffon: Quadrupedes*. 14 vols. Paris: Didot, 1799.

Lenckos, Elisabeth. "Daylesford." *The East India Company at Home, 1757–1857*. August 2014. http://blogs.ucl.ac.uk/eicah/files/2014/08/Daylesford-PDF-Final-20.08.141.pdf.

Lennox, Charlotte. *Entertaining History of the Female Quixote, or The Adventures of Arabella*. London, 1752.

Lester, Alan. "Imperial Circuits and Networks: Geographies of the British Empire." *History Compass* 4, no.1 (2006): 124–141.

———. *Imperial Networks: Creating Identities in Nineteenth-Century South-Africa and Britain.* London: Routledge, 2001.

Lettsom, John Coakley. *The Natural History of the Tea-Tree, with Observations on the Medical Qualities of Tea, and Effects of Tea-Drinking.* London: Edward and Charles Dilly, 1772.

Leung, Ka Lun. "Missions, Cultural Imperialism, and the Development of the Chinese Church." In *After Imperialism: Christian Identity in China and the Global Evangelical Movement,* edited by Richard R. Cook and David W. Pao, 23–34. Eugene, OR: Pickwick, 2011.

Linnaeus, Carl, and Pehr Cornelius Tillaeus. *Potus Theae.* Uppsala, 1765.

Liu, Lydia H. *The Clash of Empires: The Invention of China in Modern World Making.* Cambridge, MA: Harvard University Press, 2004.

London Evening Post, October 22–24, 1771.

London Gazette, June 6–10, 1706, 2–3.

London Gazette, December 3–7, 1696, 2.

London Magazine 1, January 1820.

Lowenthal, Cynthia. *Performing Identities on the Restoration Stage.* Carbondale and Edwardsville: Southern Illinois University Press, 2003.

Macartney, Sir George. *An Embassy to China: Being the Journal Kept by Lord Macartney during His Embassy to the Emperor Ch'ien-lung, 1793–1794.* Edited with introduction and notes by J. L. Cranmer-Byng. London: Longmans, 1962.

———. *The Private Correspondence of Lord Macartney Governor of Madras (1781–85).* Edited by C. Collin Davies. London: Royal Historical Society, 1950.

Mack, Maynard. *The Garden and the City: Retirement and Politics in the Later Poetry of Pope, 1731–1743.* Toronto: Toronto University Press, 1969.

Mackenzie, Henry. *The Lounger* 20 (June 18, 1785): 168–175.

Mackenzie, John M. "Imperial Networks." *Britain and the World* 11, no. 2 (2018): 149–152.

MacLean, Gerald, and Nabil Matar. *Britain and the Islamic World, 1558–1713.* Oxford: Oxford University Press, 2011.

Mair, Victor H., and Erling Hoh. *The True History of Tea.* London: Thames and Hudson, 2009.

Malton, James. *An Essay on British Cottage Architecture: Being an Attempt to Perpetuate on Principle, that Peculiar Mode of Building, which was Originally the Effect of Chance . . .* London: Hookham and Carpenter, 1798.

Markley, Robert. "Alexander Hamilton, the Mughal War, and the Critique of the East India Company." Special issue, "Transcultural Networks in the Indian Ocean, Sixteenth to Eighteenth Centuries: Europeans and Indian Ocean Societies in Interaction," edited by Su Fang Ng. *Forms of Discourse and Culture* 48, no. 2 (2015): 237–259.

———. "China and the English Enlightenment: Literature, Aesthetics, and Commerce." *Literature Compass* 11, no. 8 (2014): 517–527.

———. "China and the Making of Global Modernity." *Eighteenth-Century Studies* 43, no. 3 (2010): 299–339.

———. "Europe and East Asia in the Eighteenth Century." *Eighteenth Century: Theory and Interpretation* 45, no. 2 (2004): 111–206.

———. *The Far East and the English Imagination, 1600–1730.* Cambridge: Cambridge University Press, 2006.

———. "Gulliver and the Japanese: The Limits of the Postcolonial Past." *Modern Language Quarterly: A Journal of Literary History* 65, no. 3 (2004): 457–479.

Marschalk, Lacy. "'What Strikes the Eye': The Forgotten India Sketches of Maria Graham." *Tulsa Studies in Women's Literature* 35, no. 2 (2016): 513–520.

Marshall, John. *John Locke, Toleration and Early Enlightenment Culture.* Cambridge: Cambridge University Press, 2009.

Martin, Benjamin. *The Description and Use of an Orrery of a New Construction.* London, 1771.

———. *Philosophia Britannica.* London, 1747.

Mason, William. *The English Garden: A Poem. In Four Books.* Dublin: S. Price et al., 1782.

Matar, Nabil. "The Arabic-Speaking Peoples and 'Globalization': The Eighteenth Century." *Eighteenth Century: Theory and Interpretation* 58, no. 1 (2017): 121–125.

———. *Europe Through Arab Eyes, 1578–1727.* New York: Columbia University Press, 2009.

McKechnie, Samuel, "Charles Lamb of the India House." *Notes and Queries* 191, no. 10 (November 16, 1946): 204–206.

McKendrick, Neil, John Brewer, and J. H. Plumb. *The Birth of a Consumer Society: The Commercialization of Eighteenth-Century England*. London: Europa, 1982.

Medeiros, Michelle. "Crossing Boundaries into a World of Scientific Discoveries: Maria Graham in Nineteenth-Century Brazil." *Studies in Travel Writing* 16, no. 3 (2012): 263–285.

Mee, Jon, ed. "Networks of Improvement: Literary Clubs and Societies c. 1760–c. 1840." Special issue, *Journal for Eighteenth-Century Studies* 38, no. 4 (2015): 475–612.

Melillo, Edward D. "Empire in a Cup: Imagining Colonial Geographies through British Tea Consumption." In *Eco-Cultural Networks and the British Empire: New Views on Environmental History*, edited by James Beattie, Edward Melillo, and Emily O'Gorman, 68–91. London: Bloomsbury, 2015.

Mercurius Politicus. September 23, 1658.

Millburn, John R. "Benjamin Martin and the Development of the Orrery." *British Journal for the History of Science* 6, no. 4 (December 1973): 378–399.

Miller, Philip. *The Gardeners Dictionary: Containing the Methods of Cultivating and Improving the Kitchen, Fruit and Flower Garden; as also, the Physick Garden, Wilderness, Conservatory, and Vineyard; According to the Practice of the Most Experinc'd Gardeners of the Present Age*. London, 1731.

Millington, Mark. "Transculturation: Taking Stock." In *Transculturation: Cities, Spaces and Architectures in Latin America*, edited by Felipe Hernández, Mark Millington, and Iain Borden, 204–234. Amsterdam: Rodopi, 2005.

Milner-Thornton, Juliette Bridgette. "Imperial Networks in a Translational Context." In *The Long Shadow of the British Empire: The Ongoing Legacies of Race and Class in Zambia*, 107–129. New York: Palgrave Macmillan, 2012.

Milton, John. *Paradise Lost*. Edited by Scott Elledge. New York: W. W. Norton, 1975.

Molesworth, Jesse. "Time and the Cosmos: The Orrery in the Eighteenth-Century Imagination." Paper presented to the Early Modern Studies Group at the University of Chicago, 2016.

Molloy, Charles. *The Coquet: Or, the English Chevalier*. London: E. Curll and R. Francklin, 1718.

Monsman, Gerald. *Confessions of a Prosaic Dreamer: Charles Lamb's Art of Autobiography.* Durham, NC: Duke University Press, 1984.

Morrison, Eliza. *Memoirs of the Life and Labours of Robert Morrison.* 2 vols. London, 1839.

Morrison, Robert. *China; A Dialogue, For the Use of Schools: Being Ten Conversations between a Father and His Two Children, Concerning the History and Present State of That Country.* London, 1824.

———. *Dialogues and Detached Sentences in the Chinese Language; With a Free and Verbal Translation in English. Collected from Various Sources. Designed as an Initiatory Work for the Use of Students of Chinese.* Macao, 1816.

Mounsey, Chris. "'. . . bring her naked from her Bed, that I may ravish her before the Dotard's face, and then send his Soul to Hell': Penelope Aubin, Impious Pietist, Humourist or Purveyor of Juvenile Fantasy?" *Journal for Eighteenth-Century Studies* 26 (2003): 55–75.

———. "Conversion Panic, Circumcision, and Sexual Anxiety: Penelope Aubin's Queer Writing." In *Queer People: Negotiations and Expressions of Homosexuality, 1700–1800,* edited by Chris Mounsey and Caroline Gonda, 246–260. Lewisburg, PA: Bucknell University Press, 2007.

Mounsey, Chris, and Caroline Gonda, eds. *Queer People: Negotiations and Expressions of Homosexuality, 1700–1800.* Lewisburg, PA: Bucknell University Press, 2007.

Mulholland, James. "Connecting Eighteenth-Century India: Orientalism, Della Cruscanism, and the Translocal Poetics of William and Anna Maria Jones." In *Representing Place in British Literature and Culture, 1660–1830: From Local to Global,* edited by Evan Gottlieb and Juliet Shields, 117–136. Farnham: Ashgate, 2013.

———. *Sounding Imperial: Poetic Voice and the Politics of Empire, 1730–1820.* Baltimore, MD: Johns Hopkins University Press, 2013.

Munster, Anna. *An Anesthesia of Networks: Conjunctive Experience in Art and Technology.* Cambridge, MA: MIT Press, 2013.

Murray, Chris. "'Wonderful Nonsense': Confucianism in the British Romantic Period." *Interdisciplinary Literary Studies* 17, no. 4 (2015): 593–616.

Muthu, Sankar. "Adam Smith's Critique of International Trading Companies: Theorizing 'Globalization' in the Age of Enlightenment." *Political Theory* 36, no. 2 (2008): 185–212.

Nechtman, Tillman W. *Nabobs: Empire and Identity in Eighteenth-Century Britain*. Cambridge: Cambridge University Press, 2010.

Ng, Su Fang, ed. "Transcultural Networks in the Indian Ocean, Sixteenth to Eighteenth Centuries: Europeans and Indian Ocean Societies in Interaction." Special issue, *Forms of Discourse and Culture* 48, no. 2 (2015): 119–340.

Nicolson, Benedict. *Joseph Wright of Derby: Painter of Light*. 2 vols. London: Routledge and Kegan Paul, 1968.

Nieuhof, Johan. *An Embassy from the East-India Company of the United Provinces, to the Grand Tartar Cham Emperour of China*. Translated by John Ogilby. London: printed by John Macock for the Author, 1669.

———. *Het Gezantschap der Neêrlandtsche Oost-Indische Compagnie aan den Grooten Tartarischen Cham, den Tegenwoordigen Keizer van China*. Amsterdam, 1665.

Nussbaum, Felicity A., ed. *The Global Eighteenth Century*. Baltimore, MD: Johns Hopkins University Press, 2003.

O'Collins, Gerald. *Christology: A Biblical, Historical, and Systematic Study of Jesus*. Oxford: Oxford University Press, 2009.

Ogilvie, Marilyn B. *Searching the Stars: The Story of Caroline Herschel*. Stroud: History Press, 2011.

O'Neill, Lindsay. *The Opened Letter: Networking in the Early Modern British World*. Philadelphia: University of Pennsylvania Press, 2014.

O'Quinn, Daniel. *Entertaining Crisis in the Atlantic Imperium, 1770–1790*. Baltimore, MD: Johns Hopkins University Press, 2011.

Ortiz, Fernando. *Cuban Counterpoint: Tobacco and Sugar*. Translated by Harriet de Onís. New York: Alfred A. Knopf, 1947.

Osbeck, Pehr. *Dagbok öfwer en Ostindisk resa åren 1750, 1751, 1752*. Stockholm: Lor. Ludv. Grefing, 1757.

Osbeck, Peter. *A Voyage to China and the East Indies*. Translated from the German by John Reinhold Forster. 2 vols. London: Benjamin White, 1771.

———. *Reise nach Ostindien und China*. Edited and translated by Johann Gottlieb Georgi. Rostock: Johann Christian Koppe, 1765.

Ovington, John. *An Essay upon the Nature and Qualities of Tea*. London: R. Roberts, 1699.

————. *A Voyage to Suratt, in the Year, 1689: Giving a Large Account of That City, and Its Inhabitants, and of the English Factory There.* London: Jacob Tonson, 1696.

Oxford English Dictionary. 2nd ed. Oxford: Oxford University Press, 1989. Online edition. http://www.oed.com/.

Pachter, Lior, and Nicolas Bray. "The Network Nonsense of Albert-László Barabási." *Bits of DNA*, February 10, 2014, blog. https://liorpachter .wordpress.com/2014/02/10/the-network-nonsense-of-albert-laszlo -barabasi/.

Pagani, Catherine. *"Eastern Magnificence & European Ingenuity": Clocks of Late Imperial China.* Ann Arbor: University of Michigan Press, 2004.

Parker, Heidi G., Dayna L. Dreger, Maud Rimbault, et al. "Genomic Analyses Reveal the Influence of Geographic Origin, Migration, and Hybridization on Modern Dog Breed Development." *Cell Reports* 19, no. 4 (2017): 697–708.

Peckham, Robert. "Game of Empires: Hunting in Treaty-Port China, 1870–1940." In *Eco-Cultural Networks and the British Empire: New Views on Environmental History*, edited by James Beattie, Edward Melillo, and Emily O'Gorman, 202–232. London: Bloomsbury, 2015.

Pérau, Gabriel-Louis Calabre. *L'Ordre de Francs-Maçons Trahi et le Secret des Mopses révélé.* Amsterdam: Henri-Albert Gross & Co., 1745.

Perez-Mejia, Angela. *A Geography of Hard Times.* Translated by Dick Custer. Albany: State University of New York Press, 2004.

Petiver, James. "An Account of Mr Sam. Brown, his Third Book of East India Plants, with Their Names, Vertues, Description, &c. By James Petiver, Apothecary, and Fellow of the Royal Society. To Which are Added Some Animals Sent Him from Those Parts." *Philosophical Transactions of the Royal Society of London* 22 (1700–1701): 843–862.

————. *Catalogus Classicus et Topicus.* London: Christopher Bateman, 1709.

————. "A Description of some Coralls, and other Curious Submarines . . . as also An Account of some Plants from Chusan and Island on the Coast of China; Collected by Mr James Cuninghame, Chyrurgeon & F.R.S." *Philosophical Transactions of the Royal Society of London* 23 (1702–1703): 1419–1429.

————. *Gazophylacii Naturæ et Artis Decas Tertia.* London: Samuel Smith and Christopher Bateman, 1704.

Petiver, James, and Sam. Brown. "An Account of Mr Sam. Brown His
Fifth Book of East India Plants, with Their Names, Vertues, Descrip-
tion, &c by James Petiver, Apothecary and Fellow of the Royal
Society. To Which are Added Some Animals Sent Him from Those
Parts." *Philosophical Transactions of the Royal Society of London* 22
(1700–1701): 1007–1029.

———. "An Account of Mr Sam. Brown His Second Book of East India
Plants, with Their Names, Vertues, Description, etc. By James Petiver,
Apothecary, and Fellow of the Royal Society. *Philosophical Transactions
of the Royal Society of London* 22 (1700–1701): 699–721.

———. "An Account of Mr. Sam. Brown His Sixth Book of East India
Plants, with Their Names, Vertues, Description, etc. By James Petiver,
Apothecary, and Fellow of the Royal Society. To These are Added
Some Animals, etc. Which the Reverend Father George Joseph
Camel, Very Lately Sent Him from the Philippine Isles." *Philosophical
Transactions of the Royal Society of London* 23 (1702–1703): 1055–1068.

———. "An Account of Part of a Collection of Curious Plants and Drugs,
Lately Given to the Royal Society by the East India Company / An
Account of Mr Sam. Brown, his First Book of East India Plants, with
Their Names, Vertues, Description, &c. And some additional Remarks
by James Petiver, Apothecary, and Fellow of the Royal Society." *Philo-
sophical Transactions of the Royal Society of London* 22 (1700–1701): 579–594.

———. "A Description of Some Shells Found on the Molucca Islands; as
Also an Account of Mr Sam. Brown, his Fourth Book of East India
Plants, with Their Names, Vertues, etc. By James Petiver, Apothecary
and Fellow of the Royal Society." *Philosophical Transactions of the Royal
Society of London* 22 (1700–1701): 927–46.

———. "Mr Sam. Brown His Seventh Book of East India Plants, with an
Account of Their Names, Vertues, Description, etc. By James Petiver,
Apothecary, and Fellow of the Royal Society. These Plants Were
Gathered between the 15th and 20th of June, A.D. 1696. in the Ways
between Fort St George and Trippetee, which is about 70 Miles off."
Philosophical Transactions of the Royal Society of London 23 (1702–1703):
1251–1266.

Peyrefitte, Alain. *The Collision of Two Civilisations: The British Expedition to
China, 1792–4.* Translated by Jon Rothschild. London: Harvill, 1992.

————. *The Immobile Empire*. New York: Vintage, 1992.

Phillips, Harry. *The Valuable Library of Books in Fonthill Abbey: A Catalogue of the Magnificent, Rare and Valuable Library (of 20,000 Volumes)*. 2 vols. London: Phillips, 1823.

Piozzi, Hester Lynch. *Observations and Reflections Made in the Course of a Journey through France, Italy, and Germany*. 2 vols. London: A. Strahan and T. Cadell, 1789.

Platt, Stephen R. *Imperial Twilight: The Opium War and the End of China's Last Golden Age*. New York: Vintage, 2018.

Pope, Alexander. *The Rape of the Lock, An Heroi-Comical Poem in Five Canto's*. 4th ed. London: Bernard Lintott, 1715.

————. *The Poems of Alexander Pope: A One-Volume Edition* (1963). Edited by John Butt. New Haven, CT: Yale University Press, 1977.

Porter, David. *The Chinese Taste in Eighteenth-Century England*. Cambridge: Cambridge University Press, 2010.

————. *Ideographia: The Chinese Cypher in Early Modern Europe*. Stanford, CA: Stanford University Press, 2001.

Pratt, Mary Louise. *Imperial Eyes: Travel Writing and Transculturation*. 2nd ed. London: Routledge, 2008.

Prescott, Sarah. "Penelope Aubin and *The Doctrine of Morality*: A Reassessment of the Pious Woman Novelist." *Women's Writing* 1, no. 1 (1994): 99–112.

Purchas, Samuel. *Purchas his Pilgrimes*. 4 vols. London: William Stansby for Henrie Featherstone, 1625.

Rajan, Balachandra. *Under Western Eyes: India from Milton to Macaulay*. Durham, NC: Duke University Press, 1999.

Rawson, Claude, ed. *The Cambridge Companion to English Poets*. Cambridge: Cambridge University Press, 2011.

Repton, Humphry. *An Enquiry into the Changes of Taste in Landscape Gardening*. London: J. Taylor, 1806.

————. *Designs for the Pavillon at Brighton*. London: J. C. Stadler, 1808.

Returns showing the Number of Pounds Weight of the different Varieties of Tea sold by the East India Company, for Home Consumption, in each Year from 1740 down to the Termination of the Company's Sales, together with the Average Prices at which such Teas were sold. London: House of Commons Parliamentary Papers, 1845.

Reynolds, Joshua. *Discourses on Art.* Edited by Robert R. Wark. New Haven, CT: Yale University Press, 1975.

Rhodes, Alexandre de. *Divers Voyages et Missions du Père Alexandre de Rhodes en la Chine.* Paris: Sebastien and Gabriel Cramoisy, 1654.

Rice, Dennis G. *Dogs in English Porcelain of the 19th Century.* Woodbridge: Antique Collectors' Club, 2002.

Roebuck, Peter. ed. *Macartney of Lisanoure 1737–1806: Essays in Biography.* Belfast: Ulster Historical Foundation, 1983.

Roger, Jacques. *Buffon: A Life in Natural History.* Edited by L. Pearce Williams. Translated by Sarah Lucille Bonnefoi. Ithaca, NY: Cornell University Press, 1997.

Rolt, Richard. *New Dictionary of Trade and Commerce.* London, 1756.

Rotenberg-Schwartz, Michael, and Tara Czechowski, eds. *Global Economies, Cultural Currencies of the Eighteenth Century.* New York: AMS, 2012.

Rugg, Thomas. *The Diurnal of Thomas Rugg, 1659–1661.* Edited by William L. Sachse. Camden Third Series 91. London: Royal Historical Society, 1961.

Ryan, Dermot. *Technologies of Empire: Writing, Imagination, and the Making of Imperial Networks.* Newark: University of Delaware Press, 2013.

Saal, Wolfgang, and Frithjof Schwartz. *Der Mops: Ein Kunstwerk.* Translated by Judith Rosenhal. Trier: H2SF, 2008.

Said, Edward. *Orientalism.* New York: Pantheon, 1978.

Sainsbury, Ethel Bruce, ed. *A Calendar of the Court Minutes etc. of the East India Company, 1664–1667.* Oxford: Clarendon, 1925.

Savolainen, Peter, Ya-ping Zhang, Jing Luo, Joakim Lundeberg, and Thomas Leitner. "Genetic Evidence for an East Asian Origin of Domestic Dogs." *Science* 298, no. 5598 (2002): 1610–1613.

Schiebinger, Londa. *Plants and Empire: Colonial Bioprospecting in the Atlantic World.* Cambridge, MA: Harvard University Press, 2004.

Schlegel, G[eorge]. "First Introduction of Tea into Holland." *T'oung Pao* (Second series) 1 (1900): 468–472.

Schmidt-Haberkamp, Barbara, ed. *Europa und die Türkei im 18. Jahrhundert / Europe and Turkey in the 18th Century.* Bonn: Bonn University Press; Göttingen: V. & R. Unipress, 2011.

Sen, Amrita. "Traveling Companions: Women, Trade, and the Early East India Company." In special issue, "Transcultural Networks in the

Indian Ocean, Sixteenth to Eighteenth Centuries: Europeans and Indian Ocean Societies in Interaction," edited by Su Fang Ng. *Forms of Discourse and Culture* 48, no. 2 (2015): 193–214.

Sen, Krishna. "A Passage to India: India and the Imperial Gaze in Maria Graham's *Journal of a Residence in India*." *Nineteenth-Century Literature in English* 9, no. 1 (2005): 231–259.

Senex, John. *A New Map of India & China: From the Latest Observations* (1721). https://www.raremaps.com/gallery/detail/54181/a-new-map-of-india -china-from-the-latest-observations-senex.

Sheridan, Louisa Henrietta, ed. "An Auction for Wives." In *The Comic Offering; Or, Ladies' Melange of Literary Mirth for 1834*. London: Smith, Elder and Co., 1834.

Shih, Shu-mei. "The Concept of the Sinophone." *PMLA* 126, no. 3 (2011): 709–718.

Short, Thomas. *A Dissertation upon Tea*. London: W. Bowyer, 1730.

Sibthorpe, Jan. "Sezincote, Gloucestershire." *East India Company at Home*. August 30, 2014. http://blogs.ucl.ac.uk/eicah/files/2013/06/Sezincote -PDF-Final-19.08.14.pdf.

Smith, Adam. *The Glasgow Edition of the Works and Correspondence of Adam Smith*. Vol. 5, *Lectures on Jurisprudence*, edited by R. L. Meek, D. D. Raphael, and P. G. Stein. Oxford: Clarendon, 1978.

———. *An Inquiry into the Nature and Causes of the Wealth of Nations*. Edited by R. H. Campbell A. S. Skinner, and W. B. Todd. 2 vols. Oxford: Clarendon, 1976.

Smith, James Edward, ed. and trans. *A Selection of the Correspondence of Linnæus, and Other Naturalists*. 2 vols. London: Longman et al., 1821.

Solar, Peter M. "Opening to the East: Shipping between Europe and Asia, 1770–1830." *Journal of Economic History* 73, no. 3 (2013): 625–661.

Solkin, David. "ReWriting Shaftesbury: The Air Pump and the Limits of Commercial Humanism." In *Painting and the Politics of Culture: New Essays on British Art, 1700–1850*, edited by John Barrell, 73–99. Oxford: Oxford University Press, 1992.

"Some Account of a Hindu Temple, and a Bust, of which Elegant Engravings Are Placed in the Oriental Library of the Hon. East India Company, in Leadenhall Street. [With Two Plates.]" *European Magazine* 42 (December 1802): 448–449.

Southey, Robert. *The Curse of Kehama*. London: Longman, 1810.

Spanderi, Elena. "Beyond Fellow-Feeling? Anglo-Indian Sympathy in the Travelogues of Eliza Fay, Maria Graham, and Fanny Park." *Textus: English Studies in Italy* 25, no. 2 (2012): 127–144.

Spence, Jonathan D. *The Memory Palace of Matteo Ricci*. New York: Viking, 1984.

Spencer, Jane. *The Rise of the Woman Novelist: From Aphra Behn to Jane Austen*. Oxford: Basil Blackwell, 1986.

Staley, Lynn. *The Island Garden: England's Language of Nation from Gildas to Marvell*. Notre Dame, IN: University of Notre Dame Press, 2012.

Staunton, Sir George Leonard. *An Authentic Account of an Embassy from the King of Great Britain to the Emperor of China*. 3 vols. London, 1796 (vol. 3: map and plates), 1797 (vols. 1–2: text).

Staunton, Sir George Thomas, trans. *Ta Tsing Leu Lee: Being the Fundamental Laws, and a Selection from the Supplementary Statutes of the Penal Code* (1810). Cambridge: Cambridge University Press, 2012.

Staves, Susan. *A Literary History of Women's Writing in Britain, 1660–1789*. Cambridge: Cambridge University Press, 2010.

Stouffer, Daniel B., R. Dean Malmgren, and Luís A. Nunes Amaral. "Comment on *The Origin of Bursts and Heavy Tails in Human Dynamics*," October 25, 2005. https://arxiv.org/pdf/physics/0510216.pdf.

Swainston-Goodger, Wilhelmina. *The Pug-Dog: Its History and Origin*. London: Watmoughs, 1930.

Tague, Ingrid H. *Animal Companions: Pets and Social Change in Eighteenth-Century Britain*. University Park: Pennsylvania State University Press, 2015.

Tate, Nahum. *Panacea: A Poem upon Tea: In Two Canto's*. London: J. Roberts, 1700.

Thompson, Carl. "Earthquakes and Petticoats: Maria Graham, Geology, and Early Nineteenth-Century 'Polite' Science." *Journal of Victorian Culture* 17, no. 3 (2012): 329–346.

———. "'Only the Amblyrhyncus': Maria Graham's Scientific Editing of *Voyage of HMS Blonde* (1826/27)." *Journal of Literature and Science* 8, no. 1 (2015): 48–68.

Thomson, James. *The Complete Poetical Works of James Thomson*. Edited by J. Logie Robertson. Oxford: Oxford University Press, 1908.

Tickell, Thomas. *The Poetical Works of Thomas Tickell*. Edinburgh: Apollo, 1781.

Tillotson, Giles H. R. *The Artificial Empire: The Indian Landscape of William Hodges*. Richmond: Curzon, 2000.

Toreen, Olof. "A Voyage to Suratte." In *A Voyage to China and the East Indies*, Peter Osbeck, translated from the German by John Reinhold Forster. 2 vols. Vol. 2, 151–266. London: Benjamin White, 1771.

Trotta, Hans von. *Der Englische Garten: Eine Reise durch seine Geschichte*. Berlin: Wagenbach, 1999.

Twickenham Museum. "Alexander Pope's Willow Tree: A Weeping Legend." http://www.twickenham-museum.org.uk/detail.php?aid =287&ctid=2&cid=1.

"Upon a new Marry'd Lady's being fond of a *Dutch Mastiff*, which caus'd her Husband to be Jealous." *Diverting Post* 19 (February 24–March 3, 1705): 1.

Van Braam Houckgeest, Andreas Everardus. *Voyage de l'Ambassade de la Compagnie des Indes Orientales Hollandaises, vers l'Empereur de la Chine, dans les années 1794 et 1795, publié en français par M. L. E. Moreau de Saint-Méry. Philadelphie*. 2 vols. Philadelphia, 1797.

———. *An Authentic Account of the Embassy of the Dutch East-India Company, to the Court of the Emperor of China, in the years 1974 and 1795*. London: R. Philips, 1798.

Vanneste, Tijl. *Global Trade and Commercial Networks: Eighteenth-Century Diamond Merchants*. London: Pickering and Chatto, 2011.

Voltaire [François-Marie Arouet]. *The Age of Louis XIV* (1751). Translated by Martyn Percy Pollack. London: J. M. Dent, 1961.

VonHoldt, Bridgett M., John P. Pollinger, Kirk E. Lohmueller, et al. "Genome-wide SNP and Haplotype Analyses Reveal a Rich History Underlying Dog Domestication." *Nature* 464 (2010): 898–902.

Wallace, Mark C., and Jane Rendall, eds. *Association and Enlightenment: Scottish Clubs and Societies, 1700–1830*. Lewisburg, PA: Bucknell University Press, 2020.

Walpole, Horace. *The Yale Edition of Horace Walpole's Correspondence*. Edited by W. S. Lewis. 48 vols. New Haven, CT: Yale University Press, 1937–1983.

Warton, Joseph. "No. 26." In *The World, for the Year One Thousand Seven Hundred and Fifty Three*, edited by Edward Moore [pseud. Adam Fitz-Adam], 154–159. London: R. and J. Dodsley, 1753.

Wasserman, Stanley, and Katherine Faust. *Social Network Analysis: Methods and Applications*. Cambridge: Cambridge University Press, 1994.

Welham, Debbie. "The Particular Case of Penelope Aubin." *Journal for Eighteenth-Century Studies* 31, no. 1 (2008): 63–76.

———. "The Political Afterlife of Resentment in Penelope Aubin's *The Life and Amorous Adventures of Lucinda* (1721)." *Women's Writing* 20, no. 1 (February 2013): 49–63.

Welsch, Wolfgang. "Transculturality: The Puzzling Form of Cultures Today." In *Spaces of Cultures: City—Nation—World*, edited by Mike Featherstone and Scott Lash, 194–213. London: Sage, 1999.

Welschinger, Henri. "Le chien du duc d'Enghien." *Le Monde Illustré* 1656 (December 22, 1888): 403–406.

Wenger, Diane E. *A Country Storekeeper in Pennsylvania: Creating Economic Networks in Early America, 1790–1807*. University Park: Pennsylvania State University Press, 2013.

Wesley, John. *A Letter to a Friend, Concerning Tea*. 2nd ed. Bristol: Felix Farley, 1749.

White, Daniel E. *From Little London to Little Bengal: Religion, Print and Modernity in Early British India, 1793–1835*. Baltimore, MD: Johns Hopkins University Press, 2013.

White, Gilbert. *The Natural History and Antiquities of Selborne, in the County of Southampton*. London: B. White, 1789.

Whittle, Andrea, and André Spicer. "Is Actor Network Theory Critique?" *Organization Studies* 29, no. 4 (2008): 611–629.

Willinger, Walter, David Alderson, and John C. Doyle. "Mathematics and the Internet: A Source of Enormous Confusion and Great Potential." *Notices of the American Mathematical Society* 56, no. 5 (2009): 586–599.

Wilson, James. "Essays on the Origin and Natural History of Domestic Animals." *Quarterly Journal of Agriculture* 1 (May 1828—August 1829): 537–553 and 681–700.

Winner, Langdon. "Upon Opening the Black Box and Finding It Empty: Social Constructivism and the Philosophy of Technology." *Science, Technology, and Human Values* 18, no. 3 (1993): 362–378.

Winterbottom, Anna. *Hybrid Knowledge in the Early East India Company World*. Houndmills: Palgrave Macmillan, 2016.

[Wissett, Robert]. *A View of the Rise, Progress and Present State of the Tea Trade in Europe*. London: n.p., [1801?].

Wolff, Larry, and Marco Cipollini, eds. *The Anthropology of the Enlightenment*. Stanford, CA: Stanford University Press, 2007.

Wordsworth, William. *Complete Poetical Works with Introductions and Notes*. Edited by Thomas Hutchinson. New revised edition by Ernest de Selincourt. Oxford: Oxford University Press, 1988.

Worms, Laurence. "Senex, John (*bap.* 1678, *d.* 1740)." *Oxford Dictionary of National Biography*. Oxford: Oxford University Press, 2004. Online edition, September 23, 2004. https://www.oxforddnb.com/view/10.1093/ref:odnb/9780198614128.001.0001/odnb-9780198614128-e-25085?rskey=GqNXIb&result=2.

Wulf, Andrea. *The Brother Gardeners: Botany, Empire, and the Birth of an Obsession*. New York: Knopf, 2009.

Wulf, Andrea, and Emma Gieben-Gamal. *This Other Eden: Seven Great Gardens and 300 Years of English History*. London: Little Brown, 2005.

Yang, Chi-ming. "Culture in Miniature: Toy Dogs and Object Life." *Eighteenth-Century Fiction* 25, no. 1 (2012): 139–174.

———, ed. "Eighteenth-Century Easts and Wests." Special issue, *Eighteenth-Century Studies* 47, no. 2 (2014): 95–231.

———. *Performing China: Virtue, Commerce, and Orientalism in Eighteenth-Century England, 1660–1760*. Baltimore, MD: Johns Hopkins University Press, 2011.

Young, Robert, ed. *Untying the Text: A Post-Structuralist Reader*. London: Routledge, 1981.

Yu, Lu. *The Classic of Tea: Origins and Rituals*. Translated by Francis Ross Carpenter. Hopewell, NJ: Ecco, 1974.

Zimmer, Heinrich. *Myths and Symbols in Indian Art and Civilization*. Edited by Joseph Campbell. Princeton, NJ: Princeton University Press, 1972.

Zionkowski, Linda, and Cynthia Klekar, eds. *The Culture of the Gift in Eighteenth-Century England*. New York: Palgrave, 2009.

Zuroski, Eugenia. "Disenchanting China: Orientalism and the Aesthetics of Reason in the English Novel." *Novel: A Forum on Fiction* 38, no. 2/3 (2005): 254–271.

Zuroski Jenkins, Eugenia. *A Taste for China: English Subjectivity and the Prehistory of Orientalism*. Oxford: Oxford University Press, 2013.

Notes on Contributors

SAMARA ANNE CAHILL teaches literature, rhetoric, and composition at Blinn College, Texas. She received her BA from the University of Texas at Austin (2001) and her PhD from the University of Notre Dame (2009) before moving to Singapore to teach eighteenth-century anglophone literature at Nanyang Technological University (2009–2019). Her articles have appeared in *Assuming Gender, Green Humanities, Religion in the Age of Enlightenment,* and *Studies in Eighteenth-Century Culture.* She is the editor of the online journal *Studies in Religion and the Enlightenment (SRE)* and the book review editor of *1650–1850: Ideas, Aesthetics, and Inquiries in the Early Modern Era.* Her book *Intelligent Souls? Feminist Orientalism in Eighteenth-Century English Literature* was published in 2019.

GREG CLINGHAM is emeritus professor of English at Bucknell University, where, at different times, he also held the NEH Chair in the Humanities and the John P. Crozer Chair in English Literature. He is the author or editor of ten books and dozens of scholarly articles on Johnson, Boswell, Dryden, translation, memory, historiography, Orientalism, archives, the history of the book, and scholarly publishing, including *Johnson, Writing, and Memory.* He is presently writing a cultural history of Lady Anne Barnard at the Cape of Good Hope (1797–1802), while also working on Sir George Macartney's diplomatic papers from China, India, Russia, and the Cape of Good Hope. Between 1981 and 1993 Dr. Clingham taught at Cambridge, Fordham, and New York universities; and from 1996 to 2018, he was the director of Bucknell University Press, which published around 700 titles during his tenure, including 225 in eighteenth-century studies.

KEVIN L. COPE is Adams Professor of English Literature and a member of the comparative literature faculty at Louisiana State University. The author of *Criteria of Certainty*, of *John Locke Revisited*, and of *In and After the Beginning*, Cope has written scholarly essays on topics ranging from the early modern fascination with miracles to colossalism in modern culture. He edits the annual journal *1650–1850: Ideas, Aesthetics, and Inquiries in the Early Modern Era* and has served for over a decade as the General Editor of *ECCB: The Eighteenth-Century Current Bibliography*. A member of the National Governing Council of the American Society for University Professors, Cope is regularly referenced in publications such as the *Chronicle of Higher Education* and *Inside Higher Ed* and is a frequent guest on radio and television news and talk shows.

RICHARD COULTON is a senior lecturer in the School of English and Drama at Queen Mary University of London. He researches the literature and culture of the long eighteenth century, with a range of interests that include landscape, natural history, sociability, networks, collecting, London, digital humanities, horticulture, literature, and the history of the book. Dr. Coulton is the author of *Empire of Tea: The Asian Leaf that Conquered the World* (2015, with Markman Ellis and Matthew Mauger), and has also published on satire, the history of science, commercial nursery gardening, and book theft. He is currently pursuing collaborative projects with colleagues at the Natural History Museum in London, including a volume of essays about the early modern apothecary and collector James Petiver, and research into the life of James Cuninghame, a surgeon and botanist who traveled to China in the service of the East India Company.

BÄRBEL CZENNIA has served as associate professor of English at McNeese State University in Lake Charles, Louisiana, and as tenured senior lecturer of English literature at Georg-August-Universität Göttingen, Germany, for more than twenty-five years. In addition to a monograph and essays on the history of literary translation (of Charles Dickens, Joseph Conrad, William Congreve, and Richard Sheridan), she has published essays on eighteenth-century nation-building, the exploration of the South Pacific, meteorology, fireworks, female eccentrics, animals, punch bowls, sociability, and gardens in peer-reviewed journals and in scholarly books printed in Germany, the Netherlands, France, Britain, and the United States. Her essay collection *Celebrity: The Idiom of a Modern Era* appeared in 2013. She is currently working on eighteenth-century concepts of sustainability.

JENNIFER L. HARGRAVE is an assistant professor of English at Baylor University. She received her PhD in English from Rice University. She specializes in British Romanticism and its global entanglements. She has published in *Eighteenth-Century Studies, European Romantic Review, Nineteenth-Century Contexts: An Interdisciplinary Journal,* and *SEL Studies in English Literature 1500–1900*. Her current book project, *The Romantic Reinvention of Imperial China,* recovers the history of intellectual exchanges between the British and Chinese empires, showing how a literary examination of Anglo-Sino relations produces a new narrative of interimperial exchanges premised on intellectual curiosity as well as geopolitical gain.

STEPHANIE HOWARD-SMITH completed her PhD in English at Queen Mary University of London in 2018. Her thesis examined the role of lapdogs in the literary, visual, and material culture of eighteenth-century Britain. Her essay "Mad Dogs, Sad Dogs and the 'War against Curs' in London in 1760," published in the *Journal for Eighteenth-Century Studies* (42, no. 1), explores the reactions of dog lovers and haters to a rabies outbreak and subsequent cull. She has taught in the School of History and in the School of English and Drama at Queen Mary University of London, as well as in the Department of History at the University of York, where she is currently a research associate.

JAMES WATT teaches in the Department of English and Related Literature and the Centre for Eighteenth Century Studies at the University of York. He is the author of *Contesting the Gothic: Fiction, Genre, and Cultural Conflict, 1764–1832* (1999) and *British Orientalisms, 1759–1835* (2019), and he is currently working on a project about popular Orientalism in the Romantic period, provisionally titled *The Comedy of Difference*.

Index

Page numbers for figures are shown in italics.

East China Sea, 45

East Florida, 55

East India Company. *See* British East India Company; Dutch East India Company; Swedish East India Company

East Indies, 5, 45, 50, 98–99

Edinburgh, physic garden of, 101. *See also* gardens

Egypt, 9, 184, 246

Ekeberg, Carl Gustav, 48–50

Elephanta. *See* caves

Ellis, John, 49–51, 53–55; *Directions for Bringing over Seeds and Plants from the East-Indies . . . in a State of Vegetation*, 50

Ellora. *See* caves

embassy to China: British: first (Lord Macartney), 190–220; second (Lord Amherst), 157, 191, 195; Dutch, 69, 109, 213n1, 216–217n22

empire: anxieties of, 182; British, 5, 7, 19–20, 94, 97, 101–102, 104, 106–107, 132, 137, 139–141, 152, 155, 169–189, 199, 222; building of, 7, 16;

empires: Celestial, 191–192; China-Tartarian, 1–72; Chinese, 44, 53, 147, 155, 160; competing, 23–24; critique of, 16–18, 20–21; Eastern, 19, 22, 136, 139, 140, 148; end of, 174; "first" (in the west), 180; Ottoman, 135, 145, 148; ruins of, 176, 181; scandal of, 184; "second" (in the East), 180; Western, 21–22

Emuy, 46. *See also* Amoy

Engelhart, Henning Munch, 102

Enghien, Duc d'; (Louis Antoine de Bourbon-Conde), 79. *See also* Mohiloff

England, 37, 44, 46, 49–51, 65, 73, 78–79, 97, 99–100, 104, 107, 109–110, 112, 115–116, 132–134, 137–139, 147–148, 160, 162–164, 181, 184, 198, 223, 252, 257, 259

Enlightenment, 5, 8, 103, 160, 192, 202, 207, 224, 247, 258

epic poem, 95, 172

Erfurt, 79

Erich, August, 71

essay (as literary genre), 20, 44–45, 164, 169–172, 175, 177, 180–184, 191, 193

Europe, 7, 9, 12, 37, 42–43, 45, 51, 66–68, 70–71, 80–82, 93, 95, 100, 102, 106, 143, 146–148, 153, 164, 205–206, 209, 238, 246, 256, 259

European: accounts of tea in China, 38; aristocrats' understanding of gift-giving, 195; astronomers, 205; attempts to Anglicize gardens in India, 15, 107; attempts to Indianize gardens in England, 105–119; bioprospectors, 35; botanists, 48; ceramics, 64; Christianity and the East, 135, 136, 139; colonies, 7; consumers of Eastern goods, 76–80; consumption of tea / coffee / chocolate in early modern period, 38; demand for Chinoiseries, 14; depictions of live pugs, 66; discovery of America, 7; dog breeds, 66–67; education in Chinese language and culture, 160; entanglement with Ottomans in the Mediterranean, 138; envoys at Chinese courts, 69; export market in China, 70; fashion, 246; garden aesthetics in India, 106; gardens transformed by Chinese plants, 55; governments' independence threatened by international trading companies, 7; ideas of cultural superiority, 180; ideas of cultural superiority resisted and deconstructed, 161; imaginary and Barbary piracy, 139; imperial interests, 18, 139, 239; inability to establish tea plant in the West, 15, 49, 54–55; lapdogs, 69; limited access to China, 48; luxury, 147; merchants in China, 36; missionaries, 152; missionary ambitions inhibited in China, 156; mythology fused with Asian, 39; nations, 7; pictorial conventions, 53, 80; plant collectors in the East, 93; plants

Malacca, 17–18, 153
Malta, 254
Malton, James, 116
Man, Henry, 179, 181
Mandarin, 69, 81, 159
Manning, Thomas, 172, 174, 183
Margate (Kent), 108
Marsden, William, 155
Martin, Benjamin, 206–207, *206*, *207*, 212;
Martin, John, 110, 115, 130n108, 174, 186n23; *View of the Temple of Suryah & Fountain of Maha Dao, with a Distant View of North Side of Mansion House*, 116
Marvell, Andrew, "To His Coy Mistress," 92
Mauk-Sow-U, *54*
Maurice, Landgrave of Hesse-Kassel, 71
Medhumadha (Hindu water nymph), 107
Mediterranean (as region), 97, 132, 136, 138–139; Eastern, 9; Muslim, 148; Ottoman, 133, 139; Sea and Barbary piracy, 139
Meissen porcelain manufactory, 63, 76–81; pugs, 76–80; snuffboxes, 77, 79. *See also* porcelain
Melchet Park, 108; pavilion of, 109; temple of, 108;
Mennecy porcelain manufactory, 78. *See also* porcelain
Mercurius Politicus, 37
Mill, James, 175
Miller, John, 51, *52*
Miller, Philip, *The Gardeners Dictionary*, 98
Milne, William, 157
Milton, John, 94, 96, 222, 231; *Paradise Lost*, 94, 173–174
missionary work: in China, 16–18, 70, 152–155, 159–160, 162–163, 165; in East Asia, 36, 70; in English fiction on China, 136, 143; in English fiction on Japan, 146; in India, 100; in Macao,

17, 36; in Malacca, 17, 157; *See also* London Missionary Society
Mitra (Vedic deity), 107
Mohiloff (famous pug dog), 79. *See also* Enghien, Duc d'
Molesworth, Jesse, 209
Molloy, Charles, *The Coquet*, 75
Monsman, Gerald, 176–177, 179
Mopsorden (Order of the Pug), 77
Mopsus (famous pug dog), 71. *See also* Lipsius, Justus
Moravians, 231
Morrison, Eliza, 153–154; *Memoirs of the Life and Labours of Robert Morrison*, 153
Morrison, John, 160, 162, 164
Morrison, Mary Rebecca, 160–161
Morrison, Robert, 17, 20, 23, 152–168; on British culture, 161–62; *China; A Dialogue, For the Use of Schools*, 154–155, 160–166; on Chinese culture, 157–158, 160–161, 164, 166; on Chinese education, 164–65; on Chinese literature, 159, 160, 164–165; on Confucius, 157, 159, 163–164; on cultural exchange, 154–155, 158–160, 162, 164; *Dialogues and Detached Sentences in the Chinese Language*, 154–155, 158–160; on female footbinding, 161–162; and the East India Company, 157, 165; and the London Missionary Society, 153, 157, 162, 166; as missionary, 153–154; as scholar, 153–154; as translator, 18, 153–154, 156–157
Moses, 243
Moultrie, John, 55
Mounsey, Chris, 134
Mount San Miguel, 236
Mughal architecture, 108, 110
Munro, Sir Hector, 108
Murray, Chris, 157
Murray, Sir John, 99
Muslim: East, 135–136; invasion, 148; Mediterranean, 148; predation of Christians in Aubin novels, 136;

saint's tomb in India, 228–229; Silk Road section, 132; women in India, 239. *See also* Islam

nabob, 24, 99, 107–108, 116
Naples, 254
Napoleon (Bonaparte), 222, 244
Navarrete, Domingo Fernández de, 94
Nechtman, Tillman, 107
Negapatam (Nagapattinam), 108
network: aesthetic, 80–82, 92–93, 104–119; of art or artists, 8, 10, 115; biological / botanical / physiological, 2, 4, 73, 93, 98, 101, 105, 256; colonial, 109, 241–244; of commerce or trade, 8, 10, 15–17, 19, 21–22, 34, 64, 79, 92–93, 97, 103, 109, 119, 132–133, 136, 138, 143, 146–148, 184, 193–194, 205, 213, 241; of communication, 2, 8, 17; computer, 4; of cosmology, 19, 22, 202, 204–205, 213; cosmopolitan, 19; critics or critique(s), 4–7, 18–20, 132–133, 135; cultural, 24, 93, 194, 198, 205, 257; definitions, 1–3, 4, 7, 92, 132, 135; discursive, 117; of disease, 2; ecocultural or ecological, 92–93, 106, 108; of empire, 6, 16–17, 19–20, 170, 184, 198; enablers, 18; of evangelism or religion, 16–17, 132, 138, 140, 143–148, 152–168; geological, 247–252; global, 22, 101–102, 143–144, 146, 148, 193, 223, 247–254; horticultural, 12, 16, 92, 103; instability of, 13, 109; institutional or public, 12, 93, 100–104; of knowledge or information, 8, 10, 12, 15, 17–18, 64, 71; maintenance of, 24; misfits, 18, 21, 157; models, 1; moral, 135–138, 148; occupational or professional, 12, 20, 23, 43, 93, 99, 170, 223; oriental, 10, 12, 17, 36, 56, 117; personal / private, 12, 15, 21, 73, 79, 81, 93–100, 102, 104, 112, 114, 223, 237; political, 8, 19, 21; of reinterpretation, 80; represented in literature, 16, 139–141, 223–227, 234, 237; resistance to, 2, 152–166;

scholarly, 24, 103, 154–155; of science, 13, 101, 103, 205; skepticism or skeptics, 2, 19; social, 2, 8, 64, 223, 243–247; subversive, 80; theory, 1; transcontinental, 16; transnational, 6; transregional, 15–16; of transport, 2–5, 15, 109; underground, 18; verbal imagery for, 3–4
networked: biography, 24; geographies, 254–259; goods, 12, 23, 243; identity, 224–235, 242–244; people, 19–21, 170, 222–224
networkers (human nodes or hubs), 4, 9–10, 15, 19, 23, 132, 134–136, 138, 143–144, 146, 221, 237–244; altruistic, 17, 132, 147; cultural, 21–22; extreme, 221–267; global, 5; imperial, 20; intercultural, 13, 18, 21, 23; networking, 2–3, 7, 11, 12–13, 15, 19, 23, 48, 92–131, 143; side effects of, 5
network hubs, 4, 10, 19, 101–102, 20
network links, 21, 23, 25, 103, 114, 184, 227
network nodes 4, 9, 15, 23, 101–103, 134–136, 138, 143–144, 146, 221, 226, 237–244, 256
networks, age of, 2
network studies, 2, 8–9; network-centered approaches, 1, 8–9, 13, 17, 21, 24
network webs, 8, 10, 12, 14, 19, 23, 65, 101–103
New England, 49
New Testament. *See* Bible
New York, 51
Nicolson, Benedict, 207
Nieuhof, Johan (*also* Johannes), 36; *An Embassy from the East-India Company of the United Province . . .* , 70–72, 197
Noah's Ark, 253
novel: in Britain, 135–142, 146, 165, 184, 224; in China, 159; episodic, 16, 142; orientalist-gothic, 78
Nymphenburg porcelain manufactory, 78. *See also* porcelain

Ogilby, John, 70–71
Okumura Masanobu, 66
opium, 181; trade, 21, 154; wars, 154–55
Orient (historic term), 8–9, 64, 68, 73, 96, 227, 255–256, 259
oriental (historic term): aesthetic, 81; appearance of pug dog, 74; art, 66; assumed heritage of pug dog, 66–67; ceremonies, 197; colonial world, 252; changing definitions of, 9; commodities, 13, 73–74; deity, 118; design, 81; design fashion, 81; fruits, 256; garden, 106, 117; hub, 19; lacquerware, 74; lapdogs, 76; missions, 16; nations perceived as culturally stagnant, 160; networks, 1, 10, 17, 36, 56, 117, 257; outlooks, 247; places perceived as outlandish, 239; plants, 255; porcelain, 81; porcelain dogs, 92; sojourn, 247; spices, 255; style of writing, 204, 257; world, 241
orientalism, 10
orientalist: attraction of porcelain, 81; focus of learned societies, 102; novel, 78; in the professional sense, 169, 237
orientalized: gardens, 117; garden owners, 117; live dogs, 14; Pandemonium of Milton, 174; porcelain dogs, 14, 79–80; porcelain motifs, 64; western consumers, 10
Oriental Repertory, 54
Ortiz, Fernando, 11, 15
Osbeck, Peter (Pehr), *Dagbok öfwer en ostindisk Resa åren*, 71
Osborne, Sir John, 108
Ottoman empire. *See* empire
Overbury, Thomas, 245
Ovington, John, 43–46; *An Essay upon the Nature and Qualities of Tea*, 44–45; *A Voyage to Suratt*, 44
Oxford, physic garden of, 101. *See also* gardens

Pagani, Catherine, 203
Paine (river), 254

Paris, 1
Park, Mungo, 156
Pasteur, Louis, 1
Peacock, Thomas Love, 170, 175
Pearl River (Zhujiang) Delta, 36
Peking (*also* Pekin), 49, 69, 190, 192, 203. *See also* Beijing
Percy, Hugh, 1st Duke of Northumberland, 51
Pernambuco, 246
Persia, 106
Persian ambassador, 183
Petiver, James, 45, 47; *Gazophylacii Naturae et Artis* ("Treasury of Nature and Crafts"), 48
Peyrefitte, Alain, 196, 204
Philosophical Transactions, 47, 202
Pigou, Frederick, 54
Piozzi, Hester Lynch, 67
Platt, Stephen R., 190
Poisson, Jeanne Antoinette. *See* Pompadour, Marquise de
Polo, Marco, 164
Pompadour, Marquise de (Jeanne Antoinette Poisson; Madame de Pompadour *in text*), 73
Pompeii, 254
Pope, Alexander, 75, 95–97, 134; *The Rape of the Lock*, 75; *The Second Satire of the Second Book of Horace Paraphrased*, 97; *Windsor-Forest*, 95–96
porcelain, 63–91, 201; aesthetics, 80; animal users of, 75; of Belgium, 79; bowls, 73; of Capodimonte, 78; of Chelsea, 78; collectors in Europe, 81; of Crailsheim, 78; critics of, 65; of Dehua, 81; of Derby (-shire), 78, 204; of Duesbury (William) & Co., 78; of Du Paquier, 78; of England, 79; export market, 64, 76, 79–81; figurines, 12, 14, 63–65, 73–82; Frankenthal, 78; of Höchst, 78; Japanese, 81; of Jingdezhen, 80–81; and literature, 73–76; of Longton Hall, 78; of Lowestoft, 78; of

Ludwigsburg, 78; of Meissen, 63, 76–81; of Mennecy, 78; modelers, 67; of Nymphenburg, 78; painters, 81; of Poland, 79; of Saint-Cloud, 78; of Schrezheim, 78; and tea culture in Britain, 40–44; vases, 41; vessels, 74, 79; of Vienna, 78; of Vincennes, 78; and visual art, 75

Potter, Paulus, *A Hall Interior with a Group of Eight Dogs*, 71

Pouncy, Benjamin Thomas, *Banyan Tree after William Hodges, 1793*, 95

Pratt, Mary Louise, 155–156, 179

Protestantism (in China), 153

providentialism, 133

Pruszków (Poland), 79. *See also* porcelain

Pugin, Augustus Charles: "The South Sea House," 176

pugs, 14, 63–91; as lapdogs, 73–76; in literature, 67, 73–76; as live animals, 66–68, 73; of Meissen, 63, 76, *77*, 78–81; as porcelain figurines, 63, *64*, 78–82; in visual art, 65–66, 71–73

Pune (*modern* Poona), 251

Qianlong: emperor, 21, 63, 190–191, 194–196, 198, 203, 205–206, 208–209, 212, 217n27; emperor's letter to George III, 191; period, 64, 79; pugs, 80

Qing, 17, 153–154, 190, 194–197, 202–203, 212; cosmology of, 191, 195, 202; government code of, 156

Quintero, 245

Radcliffe, Ann, 231

Repton, Humphry, 113–114, 116–117

Restoration (period), 34, 38

Reynolds, Sir Joshua, 114–116, 198, 225

Rhodes, Alexandre de, 36–37

Ricci, Matteo, 202

Richardson, Samuel, 137

Richmond Gardens, 101. *See also* gardens

Rio de Janeiro, 234, *235*, 243; botanical garden of, 255–256. *See also* gardens

Rodgers, Richard, *The Sound of Music*, 221

Rolt, Richard, *A New Dictionary of Trade and Commerce*, 192

Rome, 118, 231

Rosa, Salvator, 118, 228, 231

Rose, Sonya O., 175

Rossi, John, 109

Rottenhammer, Hans, *Minerva with the Muses on Mount Helicon*, 71

Rowlandson, Thomas, "The South Sea House," 176

Roxburgh, William, 99

Royal Academy, 114–115, 222

Royal Botanic Garden at Kew (*modern* Royal Botanic Gardens, Kew), 51, 99, 101–103. *See also* gardens

Royal Exchange of London, 37, 176, 182

Royal Society (of London), 47, 101–103, 153

Rugg, Thomas, 37

Ruggieri, Michele, 202

Russia, 190, 195, 246

Russian: contact of Macartney, 195; delegate to China, 69

Saint Bartholomew's Day Massacre, 144

Saint-Cloud porcelain manufactory, 78. *See also* porcelain

Salsette, 227–228, 239

San Cristovaõ, 232, 234, *234*

Sandby, Paul, 65

Sant'Angelo, ruins of, 231

Santiago, 243

Saugor, 254

Saxony, 63, 76–77

Schenk, Petrus, 80

Schrezheim porcelain manufactory, 78. *See also* porcelain

Schulte Beerbühl, Margrit, 19

Scotland, 241; Highlands, 108; loch, 258; physician, 55; surgeon, 45; traveler, 69; writer, 95–96, 105–109

Senex, John, 206

Sesincot. *See* Sezincote House and Gardens

Sevajee (Shivaji Bhosale I), 239

Sezincote House and Gardens, 109–119, *109*, *111*, *112*, *113*, *116*, *119*

Shaftesbury, 3rd Earl of (Anthony Ashley Cooper), 208

Shakespeare, William, *Romeo and Juliet*, 245

Shen Nung (*also* Shennong), 36

Shiva (Hindu deity), 111

Short, Thomas, 39

Shu-mei Shih, 154

Shunzhi emperor, 36, 197

Sibthorpe, Jan, 127n85, 128n93, 128n97, 130n104

Silk Road, 132, 147–148

Sinbad, 258

Singlô, 44, 46. *See also* tea

Sion Fort, 228

slave trade, 6–7, 139, 145–146, 175, 179–80; slavery, 21, 136, 139, 146

Sloane, Hans, 46–48

Smith, Adam, 5–7, 192; *An Inquiry into the Nature and Causes of the Wealth of Nations*, 5–7

Solander, Daniel, 50

Solkin, David, 208

Solon, 243

sonnet, 5

Southey, Robert, *The Curse of Kehama*, 95; *Roderick, the Last of the Goths*, 172

South Sea Bubble (*also* Bubble of 1720), 175–176, 179, 182

South Sea Company, 175–176, 178–180

South Sea House (London), 171, 175–176, 180–182

Spanish War of Succession, 96

Spectator, The, 3–4, 40, 176, 181–182. *See also* Addison, Joseph; Steele, Richard

Staffordshire pottery, 79

Stanmore-hill (*also* Stanmore Hill), 107–109. *See also* gardens

Staunton, Sir George Leonard, 203–204; *An Authentic Account of an Embassy from the King of Great Britain to the Emperor of China*, 203

Staunton, Sir George Thomas, 153, 195, 198

Steele, Richard, 40. See also *Spectator*; *Tatler*

St. Helena, 51

Stourhead House and Gardens, 118–119. *See also* gardens

Stowe House and Gardens, 117, 119, 134. *See also* gardens

St. Paul's Cathedral (London), 174

St. Petersburg, 195

Strabo, 68

Stuart, James Francis Edward, 134

St. Vincent, Botanic Gardens of, 55. *See also* gardens

Sungum. *See* caves

Surat (*also* Suratt; Suratte), 43–44, 71

Surya (*also* Suryah; Hindu deity), 110, 112, 118–119; Temple of, 112, 113, 116, 119

Sweden, 49

Swedish East India Company, 70

Swift, Jonathan, 134; *The Journal of a Modern Lady*, 73

Switzerland, 231

Syon House (Middlesex), 51. *See also* gardens

Table Mountain (*also* Table Top Mountain *in text*), 231, 255

Tague, Ingrid, 73, 75

Tang dynasty, 66, 76

Taoism, 159, 162

Tartary, 132, 136, 143, 252; China-Tartarian Empire, 71–72

Tate, Nahum, *Panacea*, 39

Tatler, The, 120–121n10, 126n73. *See also* Addison, Joseph; Steele, Richard

Tcha, 37. *See also* tea

tea, 12–15, 34–62, *47*, *52*, *54*, 73, 80, 101, 158–159, 171, 181, 192, 235, 255; as modern global commodity, 12–15, 34–62, 101

Teer of Arjoon, 244

Temple, Richard. *See* Cobham, 1st Viscount

Temple family, 134

Tenerife, 223, 235, 246, 256, 259

Thames (river), 41, 104

Thomson, James, *Liberty*, 95–96; *Summer* (*The Seasons*), 95–96

Tickell, Thomas, 95

Tierra del Fuego, 244

Tillaeus, Pehr Cornellius, "Potus Theae" ("The Tea-Drink"), 48–49

Tokyo, 22, 194

Toland, John, 133

Tonton (famous lapdog), 73. *See also* Deffand, Marquise du

Toreen, Olof, 70; *Voyage to Suratte*, 71

Toyo Bunko Oriental Library, Tokyo, 22, 194–195, 204, 212–213, 215–216n12

Tranquebar, 102

Tranquebarian Society, 102–103; *Det Tranquebarske Selskabs Skrifter* ("Transactions of the Tranquebar Society"), 103

transcultural: adopters, 12; marriage, 144; networks, 12; success story, 82

transculturality, 11

transculturated: biography, 24; characters, 17; gardens, 15; individuals, 24; landscape garden, 106; space, 118; taste, 99; tea, 42

transculturation, 10–11, 13–15, 18, 35, 55, 63–64, 107, 117

Trincomale (*modern* Trincomalee), 226, 258

Tryon, William, 51

Tunis, 136, 143, 146

Turgot, Jacques, 195

Turkestan, 148

Turnbull, George, 208

Turner, J.M.W., 225, 253

Twickenham, 97. *See also* gardens

"Upon a new Marry'd Lady's being fond of a Dutch Mastiff, which caus'd her Husband to be Jealous," 74

Uppsala, 48–50; botanical garden of, 48. *See also* gardens

Valparaiso, 236, 255–259

Vandergucht, Gerard, 209

Vermeer, Johannes (*also* Jan), 225

Vezel-poor (*modern* Vejalpur), 105

Vichy-Chamrond, Marie Anne de. *See* Deffand, Marquise du

Victoria and Albert Museum, 80, 82

Vienna porcelain manufactory, 78. *See also* porcelain

Vincennes porcelain manufactory, 78. *See also* porcelain

Virgil, 231; *Aeneid*, 118

Vishnu (Hindu deity), 108

Voltaire (François-Marie d'Arouet), 198

Voüi, 46. *See also* tea

Walpole, Horace, 73

Ward, Francis Swain, 115

Warton, Joseph, 65

Washington, George, 243

Wason (Charles W.) Collection, Cornell University, 22, 194, 196–197, 200–201, 215–216n12

Welham, Debbie, 134–135

Welsch, Wolfgang, 11

Wesley, John, 39

West Florida, 49

Westminster Abbey, 164

White, Daniel E., 183

White, Gilbert, 69

Wilke, Joachim, 110

William III, King of England, Scotland, and Ireland (William of Orange *in text*), 98, 142

Wilson, James, 66

Winthrop, John, 206–207

Wordsworth, Dorothy, 174

Wordsworth, Mary, 174–175